A Bitter Fog
Herbicides and Human Rights
Third Edition

by Carol Van Strum

Jericho Hill

Jericho Hill Publishing

Published July, 2021 by
Jericho Hill Publishing
Alfred, New York

Jericho Hill

A Toxic Mist, Safety Last: Tests that Fail the Test, and
Chemical Safety Tests: The Doubts Remain
used by permission of Peter von Stackelberg

Cover & Book Design by Peter von Stackelberg
Front Cover: Spray Helicopter *by United States Bureau of Land Management*
Front Cover: Clearcut Near Eugene *by Calabas*

Contents

The Murder of Melyce   1

A Bitter Fog   18

Agents Orange, White and Blue   27

Inadmissible Evidence   37

Early Warning   52

Drifting in the Wind   63

Not Only Plants   82

Critical Mass   93

"Too Many Unanswered Questions"   111

A Soldier Turns to Home   123

"Those Chemicals from Vietnam"   142

For Future Reference: The Alsea Record   166

Poisons: Innocent Until Proved Guilty   197

The Five Rivers Health Study   215

"One Little Helicopter"   234

Informed Discretion   248

Appendix A   259

Appendix B   260

Appendix C   264

Appendix D   268

Appendix E   272

Appendix F   275

Appendix G   276

Bibliography and Notes   316

Dedication
for the Second Edition

For Steve

There is but one child,
and he is all of you.
There is but one child,
and she is all children...

— Robert Cotner

Dedication
for the First Edition

For my children,

Daphne, Alexey, Juris, and Benjamin Van Strum

Acknowledgements
for the Second Edition

It's illegal to kill off your missus,
Or put poison in your old man's tea,
But poison the rivers, the seas or the skies,
And poison the minds of a nation with lies,
It's all in the interest of free enterprise,
Nonetheless it is perfectly legal.

—Ewan MacColl 1976

This second edition was made possible by the tireless efforts of Peter von Stackelberg and help from Susan Hogg, Gary Hale, Jan Wroncy, Stevens Van Strum, and the growing network of community rights and environmental justice activists worldwide.

Acknowledgements
for the First Edition

This book grew out of events in my own life that acquainted me with others who have suffered the effects of herbicide poisons and in various ways have succeeded in alerting the public to such hazards. From the start, the book was intended as a tribute to them. As such, it is a "people" book, not an academic treatise; the legal and scientific issues raised are referenced to an extensive bibliography for those who wish to examine such matters more thoroughly.

Much of the book concerns a small area in coastal Oregon, simply because that is where I live. The problems it describes, however, are global, and many dedicated people from the Pacific Northwest and other areas across the nation have given generously of their time, expertise, and experience. Without them, the book would never have been written.

In particular, I wish to thank my neighbor, friend, and ex-husband, Steve Van Strum, for goading me into writing the book and encouraging me when the task seemed overwhelming. I am indebted to Richard Kennedy, Bobbie Halperin, Judy Cornell and Christy Bauman – my "village idiots" – who read countless drafts of the manuscript, and to the dear people of Five Rivers who put up with me during its preparation. To all those whose interviews are the heart of the book, include those whose statements could not, in the end, be used, I am endlessly grateful. John Noell and Bill Bitsas provided invaluable advice and criticism on complex scientific material, particularly the discussions of genetics and biochemistry, and together with Stephen Hager provided me with a tireless research service. I am equally indebted to the staff at NCAP, Georgia Hoglund of CATH, Susan Parker of CATS, and Erik Jannson of Friends of the Earth for their encouragement and extensive research, and to Bob Morris of the Interfaith Center on Corporate Responsibility and

John Jordan of the National Council of Churches for their enthusiasm and time. I also thank Congressman Jim Weaver, Jan Kay of the Arizona Daily Star, the dedicated employees of the Oregon State Library, and all those who contributed anonymously at some risk to their careers or safety.

Without the help of two special people, this book would have taken a different and considerably less significant direction. Peter von Stackelberg of the Edmonton (Alberta) Journal unstintingly shared the results of his investigation of chemical and drug testing fraud, providing the basis for the later chapters of this book. Paul Merrell, whose interview unpredictably evolved into marriage and patchwork family for both of us, deserves immeasurable credit for editing, legal research, inspiration, and sheer endurance.

My utmost respect and gratitude go to Danny Moses, my editor at Sierra Club Books, for his unflagging confidence and patience as my initial project grew over several years into a very different work, and to my copy editor, Beatrice Rosenfeld, for her keen insights and care.

Carol Van Strum
Five Rivers, Oregon
August, 1982

Foreword to the Second Edition

This remarkable book, filled with babies, children, men, women, cats, chickens, and deer in the Coast Range of Oregon being sickened or killed by a rain of aerially-sprayed herbicides, was first published in 1983. Two years earlier, new to Oregon, I had walked casually into the Northwest Coalition for Alternatives to Pesticides (NCAP) office in Eugene, asking if they could use any volunteer help while I was finishing my botany dissertation. I ended up staying for eight years as a staff scientist, working to help citizens wend their way through the toxics on one hand and blind and deaf chemical companies, industry apologists, and state and federal agencies on the other, to stop the poisonings.

But when I first walked in that door, I was unaware of what pain and defiance had led to the formation of NCAP: all the illnesses and deaths; the frustration at being repeatedly sprayed upon, without notification, from the air; the water and tissue samples that had been taken by health and federal agencies with results sequestered for months or lost forever; the refusal of the Forest Service to cease the bombardments of their constituents. I was unaware of all the unlikely characters who had emerged from their comfort zones to challenge the senseless sprayings.

So this book – it's more than 30 years old. Why read it now? Four good reasons:

1. It's all still happening now. Now, in late 2014, it's the same, seemingly idyllic Coast Range and it's the same scene of babies, children, men, women, cats, chickens, and deer suddenly sickened or dead after herbicide soups have been dumped on them. It's not 2,4,5-T plus 2,4-D sprayed by the Forest Service now; it's 2,4-D, glyphosate, picloram and dicamba and who knows what else sprayed by the private timber companies that own the Oregon coastal forests that are still being clearcut.

2. It's happening to you. *A Bitter Fog* never was about just coastal Oregon. It was about U.S. soldiers and rural Vietnamese alike, sickened and dying from the same herbicides. It was about the upside-down priorities of the Environmental Protection Agency, which remains incredibly more protective of industry than of you and me and salmon. Just recently, the EPA announced its registration of Enlist Duo, a mixture of 2,4-D and glyphosate, promising that this time around, it "will be protective of the public, agricultural workers, and non-target species, including endangered species."
3. It's about the need for you to challenge senseless destruction. You're not a resident of Oregon's Coast Range? Maybe you live in North Dakota and you can light your drinking water on fire from oil drilling. Maybe you're an organic farmer whose field has been contaminated by Monsanto's corn genetically modified to resist glyphosate. Maybe you live on Earth which is being heated by the most powerful industry the world has ever seen. *A Bitter Fog* is about two fundamental rights that you need to exercise, wherever you live: your right to know and your right to govern.
4. It's a riveting read. Just try to read this well-written, well-documented book and come away without images that will remain with you forever. The herbicide-exposed father of a baby born without a brain; the one-page letter to EPA summarizing post-spray miscarriages that prompts an emergency suspension of 2,4,5-T; the soldiers exposed to 2,4,5-T and 2,4-D who are sent to psychiatric units for their joint pains and anemia and fevers.

The author's own children were sickened by a roadside spraying operation while they were fishing and *A Bitter Fog* was born. Something worth reading, now.

Mary O'Brien
Castle Valley, Utah
November, 2014

Some circumstantial evidence is very strong,
as when you find a trout in the milk.

– Henry David Thoreau (1850)

The Murder of Melyce

Hear us speak; oh, hear what we say,
Who and where soever ye be ...
Unless you would die as we!
—Edna St. Vincent Millay
The Murder of Lidice, 1942

Oregon's coastal highway is a thin thread, squeezed between the Pacific Ocean on one side and dense rainforest on the other. Melyce Connelly drove along the coast, listening to a county commissioner speaking on the radio. His words rankled her. Her anger and resolve grew during the forty-mile drive to her home in the Coast Range, a home and sanctuary that was no longer safe for herself and her two small children. For more than a decade Melyce's friends and neighbors had been embroiled in a fight that pitted them against some of the world's most powerful corporations, the American government, and the U.S. military. At stake were their lives and the lives of their families.

In the temperate rainforest surrounding Melyce's home, blackberry vines and alder trees spring up almost overnight on untended clearings. Dense jungle quickly swallows abandoned homesteads and orchards, where only daffodils and an occasional apple tree remain amid the ferns and saplings, blooming tributes to years of human toil. Vast thickets of brush carpet the scarred earth of clearcuts and old logging roads. By the 1970s, the dioxin-tainted Agent Orange herbicides 2,4,5-T and 2,4-D had become indispensable tools for replacing such "unwanted vegetation" with plantations of Douglas fir saplings.

In 1979 the U.S. Environmental Protection Agency's (EPA) emergency ban of the herbicide 2,4,5-T sent shock waves through the timber and chem-

Image by Peter von Stackelberg
Temperate rain forest along Highway 34 in Oregon's Coastal Range

ical industries. They predicted the loss of 20,000 timber jobs and blamed marijuana growers for the ban of what they proclaimed to be a perfectly safe poison. On the radio, repeated at every news break, Lincoln County Commissioner Andy Zedwick vehemently denounced the 2,4,5-T ban, suggesting it was prompted by marijuana growers to protect their illegal crops. Echoing earlier Dow Chemical Company propaganda, the commissioner claimed that health problems attributed to herbicides were actually caused by smoking marijuana.

Melyce was raging by the time she reached home, a few acres of land hacked out of the forest. A single mother at the age of 22, Melyce clung doggedly to the log house she and her ex-husband had built themselves along Five Rivers, a tributary of the Alsea River, which wound through the Coast Range to empty into the Pacific. With the help of neighbors she plowed the land, drove in truckloads of manure, and coaxed a small paradise from the forest soil. Her garlic field paid the mortgage. Beds of herbs – sweet basil, lemon thyme, rosemary, dill, sage, parsley, and shallots – were sold fresh and chemical-free to restaurants up and down the coast, supplementing her winter income teaching exercise classes.

For herself, Melyce grew flowers. From March to November the log house basked in a sea of hollyhocks, roses, lilies, daisies, narcissus, columbine, dahlias, tulips, and daffodils. Her business card was a photo of herself, laughing under a cascade of flowers and garlic on her porch, a giant hibiscus blossom in her hair.

Her flowers were no comfort that day, however. Shortly before Commissioner Zedwick's broadcast, she had learned that the EPA had found dioxin in a neighbor's water supply, directly upstream from her home. The neighbor had lost two babies through miscarriages and had another child with multiple birth defects.

"You can't help but wonder if there's a connection," Melyce said.

With 2,4,5-T banned, the U.S. Forest Service announced it would substitute supposedly safe 2,4-D in its spray plans for that year, which included the headwaters of Ryan Creek, the source of water for Melyce's home. She and her neighbors met with the district ranger, who had them mark their water sources on his spray map. He promised those areas would not be sprayed.

Three days later Melyce woke to the sound of a helicopter spraying Ryan Creek. Within the next few days all of her young chicks and ducklings died. Her six-month old son developed persistent, bloody diarrhea. In the surrounding valley, during the next month every pregnant woman in her first trimester miscarried. Several children were hospitalized with near-fatal spinal meningitis. Alarmed by these events, the Lincoln County Health Department began a study of health problems following the spraying in the Five Rivers watershed. The EPA, however, took over the county's effort under the auspices of its Alsea Study. Publicity about the study had prompted the commissioner's remarks about marijuana growers.

On her farm, Melyce had collected the dead chicks and ducklings, carefully preserving them in her freezer in hope that she could get them analyzed some day. Still furious about Commissioner Zedwick's comments, Melyce retrieved the frozen chicks and ducklings from her freezer and drove 50 miles to the county offices in Newport. Carrying her infant son and the bag of frozen poultry, she marched unannounced into the commissioner's office and dropped the bag on his desk with a thump.

"Open it," she commanded.

As the startled commissioner peeled tin foil from the small, frozen bodies, Melyce placed her son on his desk and peeled off the child's diaper.

"Now sir," she said, "you tell me those ducklings died from smoking too much marijuana. You tell me those chicks died from smoking too much marijuana."

Fighting back tears, her voice shaking, she thrust a bloody, soiled diaper at the commissioner.

"You tell me this child has bloody shits day after day from smoking too much marijuana. Tell me to my face, Mr. Commissioner."

The next day the commissioner went on the radio again, this time with a public apology. Information had been brought to his attention, he said, that convinced him of grave health risks from herbicide exposure. For the rest of his time in office Commissioner Zedwick led a tireless campaign against the aerial spraying of herbicides in Lincoln County, joining the county medical society in sponsoring ballot measures to restrict such uses.

When the EPA took over the county's health study of her valley, Melyce accompanied researchers as they gathered samples. She also gave them the bodies of her chicks and ducklings, and was promised the results on dioxin and herbicide analyses within 90 days.

Melyce hounded the EPA for four years, getting only vague assurances the results would be reported to her when completed.

* * *

The long, difficult struggle that led up to the U.S. Environmental Protection Agency's (EPA) collection of samples from Melyce is recounted in the original edition of A Bitter Fog, published in 1983. The story did not end then; it continues to this day. For decades, collusion and cover ups involving major chemical companies and the EPA have kept people in the dark about both the toxic chemicals they are exposed to and the devastating health effects of those chemicals.

In February, 1983, as EPA completed its cancellation case about the serious health impacts of the herbicides 2,4,5-T and silvex, Dow Chemical Company and the U.S. EPA reached a secret agreement to end the cancellation

Image by Carol Van Strum
Arrival of the EPA sampling team at Five Rivers mill camp water supply (August 28, 1984)

Image by Carol Van Strum
EPA taking a soil core sample at Five Rivers mill camp water supply (August 28, 1984)

proceedings on these two chemicals and allow them back on the market. At the time the EPA took Melyce's dead birds, the agency was already engaged in a cover up that would endanger the health of the very people the agency was supposed to protect. From the start, EPA's agenda was to protect the toxic chemicals – and the profits they generated – that it was supposed to regulate.

As discussed in this book, a case against the U.S. Forest Service, EPA, and U.S. Bureau of Land Management – Merrell v. Block – was underway while A Bitter Fog was being written. In April, 1983, two weeks after the publication of A Bitter Fog, a federal court banned the use of herbicides by the Forest Service in Oregon's Siuslaw National Forest[1]. In 1984, the ban was extended to all federal lands in the Pacific Northwest by the Ninth Circuit Court of Appeals, which found that studies designed to prove the safety of these chemicals were missing, inconclusive, or scientifically invalid, making it impossible to "complete a comprehensive scientifically sound assessment of any potential hazards associated with the chemical."[2] The court went on to say, "EPA's data is partial at best, and suspect at worst, because of the testing scandals. The availability of the data of the chemical companies is also in question."[3]

Along with the invalid, fraudulent, or nonexistent safety studies, at issue in Merrell v. Block were the results of the unpublished Alsea Study for which Melyce had provided her dead chicks and ducklings. The EPA had refused to provide the results; in fact, the EPA claimed that no records existed of the study. As Mike Axline, the attorney representing Paul Merrell, Bonnie and Tony Hill, and others from the Alsea Study area, prepared for appeals and attorney fee hearings in the wake of the Merrell v. Block case, the Alsea Study took on new importance. Axline sought the results of the Alsea Study again. The EPA again said no records of the study existed.

Axline, curious and vexed by the disappearance of samples that Science magazine had called the smoking gun that would ban 2.4.5-T forever, went looking for answers[4]. He called Dr. Michael Gross at the University of Nebraska laboratory contracted to analyze the EPA's samples for dioxin.

Did the samples exist? Axline asked.

Of course the samples existed, Dr. Gross replied. He had analyzed them himself and sent the results to the EPA more than three years earlier.

Dr. Gross promptly faxed Axline the results from his analysis, titled "Table 7." The findings were horrifying. The dioxin 2,3,7,8-TCDD was found in drinking water samples, wildlife, and human tissue from a baby born without a brain. Dioxin levels in sediments and sludges were far higher than the levels that prompted the evacuation and ultimate abandonment of the community of Times Beach, Missouri.

At the time, another herbicide case – *NCAP v. J.R. Block et al* – was being heard in federal court against the U.S. Forest Service, U.S. Bureau of Land Management, and U.S. EPA. In sworn testimony representatives for these agencies said no dioxins had been found in samples taken from sprayed national forests. Table 7 was presented in court to rebut that testimony.

On the same day, Congressman Jim Weaver received a fax of Table 7. In Washington, D.C., Weaver demanded a congressional investigation into the EPA's cover-up of these "smoking gun" samples. In Oregon, local and state-wide media jumped on the issue of high dioxin levels in the state's vast national forests.

With the sample results of the Alsea Study now public, the EPA went into damage control mode. Within 24 hours the agency began its whitewash campaign. It was all a "mix up," the EPA said. The samples weren't from Oregon, but were from "somewhere in the mid-west," the agency claimed.

The following day Donald Barnes, the EPA's dioxin coordinator, said that the Table 7 results were at that moment in the process of being decoded[45]. His statement, of course, begged the question of how anyone could know there was a mix-up before the samples were decoded.

From that point, the mix-up story became an ever more tangled web of deceit and contradictions. The EPA's next announcement, that the samples identified in Table 7 were actually from Dow Chemical's Midland, Michigan plant, came as a surprise even to some EPA officials, who had sued Dow Chemical for preventing the EPA from entering its plant site to collect samples. In early 1983 those officials testified to Congress that Dow continued to refuse such access. The still unanswered question is how the EPA could have obtained samples from a site that it had not entered and was prevented from entering.

Two preposterous internal EPA "investigations" attributed the "mix up" and the failure to discover it as simply something that had "fallen through the cracks" in the EPA bureaucracy. How the entire multi-year, multi-agency, multimillion dollar 2,4,5-T cancellation proceeding could have fallen through the cracks is yet another unanswered question.

The news media, gullible and easily distracted, dropped the story, paying little further attention to the "mix up" and its cause. Nor did the media notice that when the EPA released reports on its "investigations," Dow Chemical and the EPA immediately announced Dow's withdrawal of all 2,4,5-T registrations and the EPA's simultaneous cancellations. Within the next few months, Dow settled both the EPA's lawsuit for access to the Midland plant and the Agent Orange case brought by Vietnam veterans. With this deal Dow got rid of some troublesome lawsuits and a chemical that was becoming impossible to support.

With the record of 2,4,5-T and dioxin toxicity conveniently buried, the EPA proceeded to replace – quietly and without public notice – its "no safe level" policy for dioxins with "risk assessment," a new technique that industry welcomed and some critics called "an exciting new technique in numerology."

In theory, risk assessment correlated the toxicity of a substance with the level of exposure in order to predict the casualty rate. Below a certain number of dead, it was okay to expose people to something that would kill a "negligible" number of them. The U.S. Supreme Court in its wisdom has determined that the number each poison can kill is one in a million people, which is "negligible" and therefore acceptable. (See, for example, *Industrial Union Department, AFLCIO v. American Petroleum Institute, 448 U.S., 607, 100 S.Ct.2844, 65L.Ed.2d 1010 (1980)*, which also recognized that thousands of identified carcinogens occur in the workplace alone, each of which presents workers with an additional, lawful "negligible risk" exposure.)

In practice, risk assessment involves adjusting all the numbers and assumptions used at every turn to keep deaths at one per million. EPA risk assessments reach a predetermined acceptable body count by substituting assumptions for actual data and cloaking the resultant nonsense in unintelligible algebraic formulas. Such legerdemain allows regulators to determine first that a certain number of deaths is "negligible," and then to juggle toxicity and exposure data to fit this "acceptable" model. (For a detailed, behind the scenes look at how EPA protects polluters and poisoners, see the exposé by E.G. Vallianatos, Poison Spring: the Secret History of Pollution and the EPA published in 2014.)

Discarding the very real human data on dioxin's reproductive toxicity developed in the 2,4,5-T proceedings, EPA numerologists assumed that its carcinogenic potency outweighed all other effects. Although dioxin was well-established as an exquisitely potent carcinogen in laboratory animals, the elusive nature of human cancer suited the risk assessor's purpose admirably: unlike reproductive effects and birth defects that take only nine months to become manifest, cancer takes decades to develop in humans, so pinpointing a causative agent is confounded by years of other exposures, lost records, the likelihood of subjects dying of other effects before cancer can emerge, and the appalling fact that the few human studies available excluded women and children altogether. This dearth of human data allowed risk assessors further to assume that dioxin was a far less potent carcinogen in humans than in all other animals.

For decades since burying its 1979 emergency "no safe level" action, as known sources of dioxin expanded to include all uses of chlorine throughout commerce, the EPA has developed increasingly arcane risk assessments to justify continued dioxin pollution, even as the evidence linking dioxin to cancer, reproductive harm, birth defects, and hormone imbalance became overwhelming. The result is that in more than fifty years up to the present, EPA has failed to establish a regulatory standard for a chemical that agency scientists compare to plutonium in its toxicity and persistence.

Although EPA's voluminous evidence that dioxin posed an "imminent hazard" of reproductive harm at any dose level was relegated to EPA archives and conveniently forgotten, a few people in Oregon and in Michigan did not forget. For four years, with help from citizens and environmental lawyers in Michigan and Oregon, I tried in court to get EPA's proof that the Oregon samples were mixed up and actually came from Michigan. After three years of litigation and some 34,000 pages of jumbled, often useless records, EPA finally produced their "definitive" proof: two pages allegedly from the log book of the EPA extraction lab in Missouri, where the samples were first sent.

Those two pages not only conflict with each other, they also conflict with the records of the labs to which the samples were purportedly sent for analysis, as well as with the sworn statements of the chemists who analyzed them. I documented all this with hundreds of pages of exhibits in a 42-page affidavit: records of samples analyzed two weeks before the analytical lab received them; records sending 100% of each sample to one lab, and a year later sending 100% to another lab; samples collected by EPA within Dow's plant while EPA was in court fighting Dow's refusal to let EPA enter the plant to collect them; samples analyzed with no record of source, collection or chain of custody; samples collected but never received in any lab, such as Melyce's chicks and ducklings and my four-eyed kitten, which simply and totally disappeared without a trace.

Recognizing after four years that EPA was unable to provide any proof of its mixup story, we ended the litigation. With 2,4,5-T and silvex no longer made or sold, the issue of dioxin in our forests or in other herbicides faded from public view.

In 1984, EPA researchers returned to Five Rivers valley to resample a single site, the water supply of Melyce's neighbor, where dioxin had been found in 1979. In the five years since 2,4,5-T was banned and 2,4-D substituted for it, dioxin levels had increased four-fold in sediments directly upstream from Melyce's home. Despite the increase, to the highest dioxin levels in stream sediments ever reported in the Pacific Northwest, the EPA announced that the levels found presented no "immediate" health risk and made no effort to collect further samples in the valley or to test the fish in Five Rivers, a major spawning stream for endangered salmon. The unfinished Alsea Study with all its data confirming the presence and effects of dioxin-tainted products simply disappeared, never to be published, its scientists subjected to a government gag order and prevented from discussing or publicizing their work.

A decade later, thirteen years after requiring manufacturers to test 2,4-D products for dioxin, documents leaked from EPA revealed that 2,4-D, sprayed over Melyce's water supply with assurances it was dioxin-free, was actually contaminated with all forms of dioxin, and one quarter of all the samples contained hazardous levels of the most toxic form, 2,3,7,8-TCDD. My Freedom of Information Act request for the dioxin analyses submitted by 2,4-D manufacturers was answered by four large boxes of blank paper stamped "Trade Secret, Confidential Business Information." Dr. Anthony Colluci, a former EPA official, pointed out that EPA had known about TCDD in 2,4-D by the early 1970s.

The use of 2,4-D in forestry and on residential lawns, roadsides, golf courses, sports fields and school grounds continues to this day, with EPA approval. Now, in a stunning repeat performance, EPA has approved Dow Chemical's new herbicide, "Enlist Duo," combining 2,4-D and glyphosate (Roundup©) to treat corn and soybeans genetically modified to withstand both poisons for weed control. Enlist Duo will increase 2,4-D use exponentially and taint all products containing soy or corn with both glyphosate and 2,4-D – with absolutely no testing of possible synergistic effects of the two poisons, to say nothing of dioxin contaminants. The fact that both 2,4-D and glyphosate were initially registered on the basis of invalid/fraudulent safety testing by IBT has been conveniently and totally forgotten.

Although the Forest Service no longer blankets entire national forests with helicopters spraying poisons, the Bureau of Land Management has substituted noxious weeds instead of timber for chemical targets. The timber industry continues to spray with reckless abandon on vast tracts of corporate-owned forests, and state and county road departments continue to spray thousands of miles of roadsides with many of the same chemicals at far higher concentrations than are used aerially. The results are sadly predictable. Within the last few years, 100% of the blood samples taken from men, women, and children in the Triangle Lake area of Oregon contained both 2,4-D and atrazine, used liberally by the timber companies. In southern Oregon an entire community was recently poisoned by helicopters spraying 2,4-D and triclopyr for timber companies.

Despite such discouraging events, there are some hopeful signs of sanity lately, particularly since the advent of the Internet has allowed rapid communication among far-flung people and greater public awareness than we dreamed of in 1976. In Oregon, initiatives proclaiming the right of communities to sustainable food and other systems are underway in three counties, and a growing community rights network is fighting state laws that shield timber companies and commercial farms from lawsuits by the neighbors and communities they poison. Both Colorado and Oregon are voting this year on measures to require labeling of food containing genetically modified ingredients. Nationwide, people increasingly oppose fracking, genetically modified foods, nanotechnology, nuclear waste transport, and other hazards, focusing on the secrecy that protects industry profits. In many cases, the same players – Dow and Monsanto are prime examples – are killing people decade after decade (within acceptable limits, of course).

Industry, with its unlimited funds, will fight these movements with all the dirty tricks in its arsenal. Scientists like Tyrone Hayes, whose research confirms the deadly nature of pesticides, will endure vicious rumor campaigns, withdrawal of research funds and grants, or, like the Alsea Study authors, be prevented from publishing their results. Citizens who fight corporate poisoning will face ugly rumors, job threats, false criminal charges and other harassments, even death threats, while industry broadcasts its lies wholesale through the media and lobbies legislators relentlessly.

* * *

A particularly effective tactic devised by the American Legislative Exchange Council (ALEC) is to craft model legislation favoring corporate donors, such as its "Right to Farm" laws that have been adopted in all 50 states[6]; the 1993 Oregon version denies poisoned individuals the right to sue the poisoners for damages and pre-empts any county or municipality attempt to protect its residents from forest spraying[6]. Thus, when Lincoln County voters in 2017 passed a measure banning aerial pesticide spraying in their county, timber and pesticide interests immediately filed suit against the county, claiming the ordinance was pre-empted by the state's right to farm law[7]. In 2019, the court ruled against the county , but for two and a half years, county residents enjoyed a total respite from aerial poison spraying[8], and because the ordinance also granted natural ecosystems the right to defend themselves in court, during that time Nature itself briefly enjoyed the right to exist unharmed[9].

Image by Andy Yardy

Lorax in an Oregon clearcut

The growing movement for both communities and natural systems to defend themselves reflects the total failure of government pesticide regulation. More than sixty years after Rachel Carson exposed the dangers of DDT, pesticides containing DDT are still on the market, as are dioxin-tainted products decades after the horrors of Agent Orange came to light; and 45 years after the discovery of massive lab fraud scandals, hundreds of pesticide registrations and food tolerances continue to be based on secret, inadequate, fraudulent, or nonexistent safety data. One such poison, glyphosate, is still sprayed aerially over our forests and water supplies despite thousands of lawsuits linking it to non-Hodgkins lymphoma. If democracy is to have any meaning left, people and Nature herself should have the right to say no to poisoning for profit.

Granting Nature the right to defend itself is not a new concept. Fifty-six years ago, Dr. Seuss wrote the parable of the Lorax, who speaks for the trees, trying to save them from an insatiable factory that consumes every tree and poisons the river, air, and soil of a once-verdant forest. A year later, Supreme Court Justice William O. Douglas eloquently argued for "the conferral of standing upon environmental objects to sue for their own preservation." Because corporations, ships, counties, and other inanimate objects have standing -- the right to be heard in court -- Justice Douglas urged that natural objects such as rivers and lakes should have the same right. "The river, for example, is the living symbol of all the life it sustains or nourishes — fish, aquatic insects, water ouzels, otter, fisher, deer, elk, bear and all other animals, including man....The voice of the inanimate object, therefore, should not be stilled."

Neither the Lorax nor Justice Douglas were heeded, and countless ecosystems have since been lost, dooming innumerable species including our own into extinction.

* * *

Giving human rights to corporations, says lawyer Gerry Spence, is like "giving an ant and a bulldozer equal rights to run over each other."[10] It's easy to feel like that ant when challenging a faceless industry behemoth, but while one ant may be powerless, a thousand ants can disable even a bulldozer, chewing through hydraulic and fuel lines, jamming control switches, swarming the hapless operator. Similarly, enough informed people can bring the machinery of death to a halt. The burgeoning community rights movement offers hope for such collective power. An informed community might encourage its grand jury and prosecutor to take the Supreme Court's grant of personhood to corporations to its logical next step: charging with premeditated murder corporate "persons" who knowingly make or use a product that kills one in a million people (which corporate-controlled regulators have allowed as "negligible," making them accessories to murder).

Yes, murder.

"From what you read and hear nowadays," wrote Agatha Christie prophetically in 1969, "it seems that murder under certain aspects is slowly but surely being made acceptable to a large section of the community."[11] Risk assessment, which defines a certain number of poisoning deaths as "negligible" and thereby acceptable, raises Christie's fiction to grim reality. Countless deaths from pesticides, workplace poisons, radioactive materials, and ubiquitous environmental pollutants are now considered acceptable public costs of corporate profit, with neither public understanding nor consent.

Risk assessment promotes the misleading fiction that the cancer death risk to an individual is one in a million, but this is false; the calculation actually estimates one cancer death for each million exposed persons. For the exposed individual dying of cancer, there is no risk, just fatal certainty. The person who spiked a few Tylenol bottles with cyanide, indifferent to the identity of its victims, was vilified as a murderer after a few people died, but risk assessment allows the same number of unidentified people to die from pesticide exposure so long as the victims are nameless and the poisoner makes a profit.

The manufacturers, users, and regulators of these poisons know as certainly as did the Tylenol poisoner that some people will die when exposed, making them equally guilty of premeditated murder. Companies whose products and wastes have poisoned entire populations for decades are in fact serial killers beyond the grotesque dreams of Ted Bundy or Jeffrey Dahmer.

As "persons" under the law, corporations can be tried for murder, and though the corporation itself cannot be put in jail, its officers can, and the corporation can be dissolved, its charter canceled, its assets seized. This will only happen when risk assessment's "negligible" form of murder is no longer acceptable to society.

* * *

On July 4, 1989, ten years after Ryan Creek was sprayed with 2,4-D, Melyce Connelly died at age 32 of brain, lung, and breast cancer. Friends and neighbors gathered in Melyce's gardens for the last time to spread her ashes among the flowers and trees she loved. Shortly thereafter, the new owners of the property bulldozed the gardens and garlic fields, and the house she had built burned to the ground. Berry vines and alder saplings now thrive in the clearing where her house and gardens once stood, the old pathways emerging ghost-like every spring in rows of bobbing daffodils.

To the friends and family and children who loved her, Melyce's death will never be negligible. A society that condones such murder cannot long endure.

Image by Scott Blackman

Melyce Connelly circa 1982

A Bitter Fog

Then a strange blight crept over the area and everything began to change. Some evil spell had settled on the community: mysterious maladies swept the flocks of chickens: the cattle and sheep sickened and died. Everywhere was a shadow of death. The farmers spoke of much illness among their families. In the town the doctors had become more and more puzzled by new kinds of sickness appearing among their patients ...

A grim specter has crept upon us almost unnoticed, and this imagined tragedy may easily become a stark reality we all shall know.

—Rachel Carson

On a spring morning in 1975, thirteen years after the publication of Silent Spring, four children went fishing in the river bordering their home in the wilderness of the central Oregon coast. The river was the pulse and breathe of their young lives. They knew it well in all its moods, and were as much at home along its shores and depths as any wild creature there. They knew the nests of thrushes, sparrows, and kingfishers along its banks, the dance ground of the grouse, the tracks of beaver, raccoon, heron, and bear on its sands, the favorite rocks of the dipper. They swam in its pools in summer and cheered the salmon up its flooded waters in winter. Log trucks and other mill traffic rumbled by on the road above the river bank, but the river world had long adapted to the noise, and the children no longer noticed it.

On this particular morning, however, a heavy tank truck crawled ponderously along the road, its engine whining in low gear. Two men perched on the back, each holding the nozzle of a thick hose that sprayed a bitter substance onto trees, brush, thistles, and ferns along the roadside. Where the road skirted the riverbank, overhanging shore and water, they directed their hoses into the water, inadvertently spraying the four children fishing down below.

The truck moved on, leaving the children gasping in a wet mist that clung to their skin and clothing. With smarting skin, tearing eyes, burning mouths, throats, and noses, they stumbled home. By nightfall, all four were sick.

In the garden close to the children's home, many plants died, their leaves twisting and wilting in grotesque configurations. During the weeks that followed, chicks, geese, and ducklings hatched on their farm deformed and unfit for life, with misshapen wings, clubfeet, crossed beaks. The family dog developed oozing sores that defied treatment, and his hind legs became paralyzed.

Within a month, Forest Service helicopters flew along nearby ridges, spraying the same bitter substance into the prevailing wind that swept the farm. The nightmare was repeated. Further up the valley, other families suffered, perplexed and helpless. Horses lost their hair, children sickened, women miscarried, cows aborted, poultry died in convulsions, garden crops and orchard trees withered.

The grim specter had passed by, and Rachel Carson's imagined tragedy had become, unopposed, a stark reality.

The four children sprayed by a county road crew that June morning in 1975 were my own. The river they were fishing in was Five Rivers, bordering our farm. The questions raised by the people of this small, remote valley—in their county and state legislatures, in federal court, and in administrative hearings in Washington, D.C.—remain largely unanswered. At the heart of their questions is the unresolved conflict between individual rights to security from poisons, and corporate or governmental rights to the "benefits" gained by poisoning others without their knowledge or consent. This conflict is one of the most compelling issues facing our technological society.

Events in our valley have not been unique. Along county roads and state highways throughout the United States, spray trucks pump the same measure of devastation. Helicopters range our national forests, parks, and timberland,

Image by Ken Gagne

Five Rivers in Oregon's Coastal Range

strewing bitter fog into the wind to settle where it will. Crop dusters swoop over vast fields of wheat, rice, and corn to spray their acrid rain. Wherever people or animals live and work in the path of these grim fogs, the same symptoms are reported again and again: in New Zealand, Sweden, and England, as well as in the United States. Often the specter passes unnoticed, for the victims do not connect their suffering with the spray; when they do notice and raise questions, they are invariably told the same thing: "The spray is harmless. It's an herbicide, it only affects plants. It's harmless to humans or animals."

Vietnamese farmers could have told them differently, having suffered the devastation not only of their crops, but also of their animals' and their own health when the same poisons were sprayed by U.S. forces in Vietnam. So also could workers in chemical industries in the United States, as well as farm laborers, tree planters, loggers, and workers hired to apply the chemicals. Scientists testing the poisons in their laboratories could have reported similar symptoms in test animals.

What my children were sprayed with, and what helicopters, crop dusters, and tank trucks continue to pump onto our food crops, forests, pastures, and roadsides, are herbicides, or plant poisons. These are the same "magic" chemicals added to lawn fertilizers to kill dandelions and crabgrass. They are dumped into ponds, irrigation ditches, lakes, and waterways to control water weeds that thrive on fertilizer, sewage, and other pollutants. As a quick, economical, and effective method of killing unwanted plants, herbicides such as the phenoxy acids, 2,4-D and 2,4,5-T, revolutionized modern agriculture.

* * *

We had moved to Five Rivers in 1974, the year before the spraying, from a small homestead in northern California. The land we chose in the valley was an old farm, surrounded on three sides by the Siuslaw National Forest. The nearest house was half a mile away. With the exception of a young farmer eight miles up the road who brought us a basket of broccoli and potatoes the day we moved in, we got to know none of our neighbors that first year. When the children were sprayed, we had no one to turn to for advice or support.

The children did no more fishing that spring day in 1975. They were too upset, and their eyes, noses, and mouths burned. I brought them to the house and washed them, changed their clothes, gave them soda to rinse their mouths, but the bitter taste and burning wouldn't go away. By evening, all were sick with nausea, vomiting, cramps, headaches, diarrhea. One of the boys had an uncontrollable bloody nose. From running through the brush to bring the kids home, I had gotten the spray on me and was sick with the same symptoms.

The children's father had run after the spray truck that morning to find out what was sprayed. The men on the truck had told him the spray was perfectly harmless.

"Just 2,4,5-T," they said. It only kills plants, and you can drink the stuff without being hurt. It wouldn't hurt the kids at all.

"We get it all over us whenever we spray and it doesn't do a thing," they said. So we told ourselves it was all a coincidence, that we'd just come down with flu or something. We had never heard of 2,4,5-T before.

It was harder to regard what happened to the river as a coincidence. The river had become a central part of our lives, and the children spent many hours there every day, fishing, swimming, exploring, observing life along the riverbank.

The day after the spray truck came, we went back to the river and saw that it had changed. The leaves of alder and willow on both sides of the river wilted, their new growth twisted in unnatural spirals. The water had an oily scum on it. Floating or beached on the shore were dead crayfish and trout, and the sodden bodies of two merganser ducklings. Caddis larvae and snails that usually traced patterns in the silt on the river bottom were dead. Songbird nests along the shore were empty, except for one, in which a hermit thrush still sat, dead.

Staring down at the dark water, we realized our sickness was no coincidence. What had happened to the river had happened—was happening—inside our children's bodies.

Our first thought was that some mistake had been made. We called the county road department, but they assured us that the mixture in the tank had been no more than 2,4,5-T. It couldn't possibly have killed any wildlife or made us sick, they said. We called the Fish and Game Department and were told the same story. Again, we tried to reassure ourselves that it was all a coincidence, that the headaches and nausea and bloody noses were caused by dust from the road or the weather or distress.

In the days following, foliage further back from the roadside began to wither and twist. The leaves of bigleaf maples along the shore turned bright yellow. Yellow patches appeared in the foliage of maples far up the steep slope across the river. In our garden, the leaves of young tomato, bean, and sunflower plants wilted, their stems spiraled like the alder twigs.

A few weeks later, the spells of sickness were dwindling. We had replanted the portions of the garden that had been killed and probably would soon have forgotten the whole incident. But on a nearby ridge one morning, a helicopter began droning back and forth. We thought it was a fire watch, until our farmer acquaintance stopped by and said the Forest Service was spraying something onto the forest. He was concerned about livestock being contaminated by whatever they were spraying.

By that time, the wind that sweeps this valley every day had brought the acrid smell of the spray to us. The children complained again of burning eyes and mouths. We called the Forest Service and asked what they were spraying. They told us it was nothing to worry about. They were spraying with 2,4-D, picloram, and 2,4,5-T to kill weed trees and make the firs grow better. It all sounded very simple and sensible. But we had just had an unpleasant experience with 2,4,5-T. Was it safe? we asked. Our livestock grazed up there on the ridge, and our children played and hiked and rode ponies up there.

"Well, perhaps you should contain your livestock," we were told, "And don't let the children play in the parts of the forest that are sprayed."

We asked how the spray could hurt them if it was so safe. The reply was evasive. "The chemicals are registered with the Environmental Protection Agency. Of course they're safe. But you should avoid contact with them as a precaution."

We stayed away from the ridge and kept the animals close to home. The herbicide was subject to no such restraint. It drifted with the wind and settled with the dew in the evening. There was no avoiding contact with it, and the smell was everywhere. Within 24 hours we were sick again, with the same symptoms as before. This time, there was a further complication, uterine bleeding in both myself and my daughter, who had not yet begun to menstruate at that time. This bleeding, along with the nosebleeds and other symptoms, recurred erratically for many months.

The replanted portions of the garden again sickened and died. In the next weeks we had an epidemic of deformities among the chicks, ducklings, and goslings that hatched on the farm: crossed beaks, stunted or missing wings, toes or whole feet on backward, and stunted growth. Very few of them survived. We tried to operate on some. An operation on one gosling with clubfeet was partially successful. One foot healed in a normal position, but the other remained twisted almost completely backward. He lived for several years. We called him Lord Byron.

We knew nothing at the time about herbicides. We were troubled, though, because such defects had never occurred in our many years of raising poultry.

During the summer following the roadside and forest spraying, the dog that had been in our family for sixteen years developed a large, oozing lump on his back. The vet couldn't explain it. He removed it, but within a few weeks similar lumps appeared all over the dog's body, particularly in his groin. His rear legs became progressively paralyzed. The vet was unwilling to remove the lumps because of the dog's age, and it was apparent the problem would only recur. By the end of the year we had to carry him in and out of the house. One night his kidneys failed. He was in such pain it was out of the question to drive him 50 miles to the vet. Shooting him was a grim kindness.

Only much later did we suspect the dog's strange disease might be connected with the spraying. He had been with the children when the county truck sprayed the river.

In trying to alert the health department, the fish and game people, the Forest Service, and the Environmental Protection Agency (EPA) to what was happening, we met only one person who truly listened to us. This was Lincoln County Commissioner Jack Postle, who was in charge of the county road department. When we told him what had happened to our kids, to the river life and the garden after the road crew sprayed, he said his father had been telling him for years that the spray was toxic to wildlife. His own hive of bees had all died mysteriously after his roadside was sprayed.

Shortly after our conversation with him, we learned that Mr. Postle explored the Alsea River (which Five Rivers flows into) and discovered for himself whole beds of dead crayfish and other dead creatures. He didn't wait for any studies or official documents to confirm his own observations, but put an immediate halt to any herbicide use by county road crews. Mr. Postle was the only official from county, state, or federal government who took our problem seriously.

Except for Mr. Postle, no officials acted. Our letters to EPA were referred to the U.S. Department of Agriculture (USDA). No studies, they answered, had ever shown the herbicides to have such effects as we had noted. The herbicides were an essential forestry tool. We could rest assured the government wouldn't dream of spraying us with something harmful.

My husband Steve spent several days at the Oregon State University library, reading studies on 2,4-D and 2,4,5-T, the phenoxy herbicides. The studies he found did not agree with the USDA statements, much as we wanted to believe them. He found many studies and reports that documented effects of herbicides similar to what had happened on our farm and river. We wrote more letters – to congressmen, senators, state officials, EPA – telling them our story, enclosing copies of the studies Steve had found. Some of them referred our letters to USDA; others reiterated USDA statements. Official lack of concern bewildered us. We contacted other people for more information. One was a biologist friend, who had once worked for the Forest Service. We asked him if our problems could be related to the spraying of the herbicides.

"They can't do that!" he said. "Do you know what that stuff is? It's an analogue to growth hormones, it affects cells – any cells, plant or animal. It destroys the immune system. It screws up genetic mechanisms. That means it can cause cancer and mutations. They had to stop spraying it in Vietnam because of what it did to the people over there."

We had called him hoping he would tell us we were mistaken. Instead, he sent us photocopies of all the studies he could find in the library of the University of California at Berkeley and a list of other references.

It was fall by this time. The Forest Service had issued a new environmental impact statement (EIS) for their 1976 herbicide program. We had trouble convincing them to give us a copy. At first they said that the copy they had sent to the Newport Library – 46 miles away – was accessible enough to us. Not one of the studies our friend had sent us was referred to in the EIS.

About this time, we wrote another letter that brought an unexpected response. This was to Thomas Whiteside of The New Yorker. We remembered his articles from 1969 and 1970 that later became the book, Withering Rain, about the use and effects of Agent Orange in Vietnam, but couldn't recall much detail from the book. Since neither the university library nor the public library had a copy of the book, we wrote a brief note to Whiteside, describing our situation and asking for recent information on phenoxy herbicides. Three days later, he telephoned. He listened to our story with interest and answered some of our questions about the extent of herbicide use. He said he felt very strongly that serious doubts had been raised by responsible scientists about the safety of 2,4,5-T. Its continued use without really satisfactory proof of its safety was unjustified, he thought, particularly in programs directed by an agency of the federal government.

In reply to our request for recent information about the effects of 2,4,5-T, Mr. Whiteside suggested several scientific papers by scientists who, he said, had shown particular concern about the potential hazards of 2,4,5-T to public health. When we wrote to these scientists, some of them were most responsive, both in explaining their own work and in suggesting further research.

In contrast to the official indifference we had encountered, many responsible scientists shared our concern about the poisons being sprayed on us. Hoping to add other voices to our own, we wrote to environmental groups and publications. Only one responded—indirectly. Audubon magazine had referred our letter to the Dow Chemical Company. Dow sent a thick package of company literature, including an elegant brochure on phenoxy herbicides. The brochure was illustrated lavishly with color nature photographs by Eugene Kenaga, an associate scientist for Dow who was also president of the Michigan Audubon Society.

The cover letter, explaining that Audubon had referred our letter to Dow for reply, was signed by Dr. Kenaga. He praised us for our concern and repeated what we had already heard from USDA. The herbicides were safe. No harm could result if the EPA label instructions were followed.

After this response from Audubon/Dow, our hopes for awakening any official or public concern over herbicides were exhausted. We wrote a response to the draft of the Forest Service EIS in the fall of 1975, including with it a copy of an EPA memo from August, reporting levels of a highly toxic contaminant of 2,4,5-T called TCDD in the fat of cattle grazed on herbicide-treated pastures. Then we put all of our papers away, feeling at the time that we could do no more.

Agents Orange, White and Blue

The idea of killing plants with poisons originated more in military than in agricultural history. Heavy doses of salt are an ancient, effective method of killing plants, poisoning the soil for future use indefinitely. In 300 B.C., the Romans salted the croplands of Carthage, intending the land to "lie desolate forever." Weed control, however, requires more selective poisons that affect only weeds and leave crop plants unaffected.

Prior to World War II, weed-control methods using table salt, sodium nitrate, iron sulfate, copper salts, and even sulfuric acid offered little improvement over Roman methods of destroying crops at Carthage.

Sophisticated, "selective" methods of killing weeds with chemical poisons are a recent development in the history of agriculture. The most effective modern herbicides—phenoxy herbicides—were developed from research into methods not of killing plants, but of stimulating their growth[1]. Growth processes in plants have always been a mystery. Without brain, nervous system, or muscles, plants can "perceive" their environment and move in response to it. Young shoots bend toward the sunlight, flowers bloom at appointed times, vine stems and tendrils seek support and climb upwards. In 1880, Charles Darwin proposed that plant growth is regulated by some hormone-like chemical, which he called auxin. In 1940, researchers succeeded in isolating indoleacetic acid (IAA), the plant-growth-regulating hormone or auxin, hypothesized by Darwin sixty years earlier. Attempts to synthesize similar plant-growth-regulating compounds led to the development of phenoxy acids, 2,4-D and 2,4,5-T. Extremely small quantities of these synthetic plant hormones could stimulate cuttings to root, tomatoes to grow without seeds, fruit to ripen uniformly.

Researchers quickly discovered that slight excesses of these exciting new chemicals could kill plants instead of stimulating their growth. Most significantly, overdoses of the compounds had a selective effect, killing some kinds of plants but leaving others relatively unaffected. This selective lethal quality of the chemicals attracted keen interest from scientists in another field. Within a year after the isolation of indoleacetic acid, the potential use of synthetic growth-regulating hormones to kill plants was being studied by Dr. E.J. Kraus, head of the botany department at the University of Chicago, which had a contract with the U. S. Army for research in chemical warfare.

In 1943, Kraus presented a formal report on the crop-killing potential of synthetic plant hormones – including 2,4-D and 2,4,5-T – to a National Academy of Sciences committee on biological warfare. The following year, after the U.S. Army's contract with the University of Chicago ended, Kraus transferred to Camp Dietrick, the Army's biological warfare testing center in Maryland. Over the next year and a half he directed the testing of over a thousand chemical agents on plants. In the hands of the military, the stuff of life had become an agent of death.

Some of the earliest research on both the life-promoting and death-dealing potentials of phenoxy compounds was performed by two former students of Kraus at the U.S. Department of Agriculture's research station at Beltsville, Maryland, close neighbor to Fort Dietrick. During the time Kraus was testing for the military, one of his former students gave some 2,4-D to a botanist employed by the U.S. Golf Association, who used it to experiment with weed control on golf courses. Her experiments led to demonstrations in 1945 of 2,4-D's effectiveness in controlling dandelions on the turf of the Mall in Washington, D.C.[2] The success of 2,4-D in controlling broadleaf weeds inspired commercial enthusiasm for the new technology. By 1980, over 70 million pounds of 2,4-D (active ingredient) were distributed domestically, and by 1981, over 1500 products containing 2,4-D were registered by the EPA for use in the United States.[3]

In 1948, some of the first experiments with phenoxy herbicides in forest management were conducted by the USDA Forest Service and Oregon State College on land near Five Rivers in the Siuslaw National Forest. These experiments demonstrated widespread drift of the aerially applied herbicides, and researchers noted substantial damage to the crop trees at all concentrations of 2,4-D[4]. Military development of herbicides continued, and by 1955 at the latest, the USDA was actively involved in U.S. Army chemical and biological warfare research on anti-crop agents[5].

In 1962, the year Rachel Carson's Silent Spring was published, Air Force "Operation Hades" propjets were flying their earliest missions in Vietnam, spraying herbicides on jungle and croplands to eliminate enemy cover and food supplies[6]. Perimeters of U. S. camps and roadsides were routinely sprayed as well, chronically exposing both American GI's and the Vietnamese population to chemicals in their air, water, and food[7]. Herbicides were shipped

Image by U.S. Air Force

U.S. military aircraft spraying herbicides over Vietnam

in color-coded barrels and were referred to by their color. Agent White was a combination of 2,4-D and picloram, the latter one of the most persistent herbicides known, considered by biologist Dr. Arthur Galston of Yale University to be "A herbicidal analogue of DDT."[8] Agent Blue was a 54 percent arsenic solution of cacodylic acid, deadly to all life. It was used to kill rice.

Agent Orange, the most widely used defoliant in Vietnam, was a fifty-fifty mixture of 2,4-D and 2,4,5-T. This was the same mixture most commonly used as a forestry herbicide in this country.

The two chemicals that comprise Agent Orange—2,4-D and 2,4,5-T— kill plants by disrupting basic growth processes. Except for an added chlorine atom in 2,4,5-T, the two compounds are alike. They differ somewhat from each other in their chemical and biological properties, their selectivity (each is more effective against different plant species), and in their production methods and costs. Both belong to the class of compounds known as phenoxy herbicides, derived from the synthesis of chlorine and phenol. Used together, they are capable of killing a wide range of broadleaf plants.

Since their development in the late 1940s, phenoxy herbicides became the focus of many scientific studies because of severe health problems in workers involved with their manufacture and application[9]. A contaminant found in 2,4,5-T was thought at first to be responsible for the herbicide's toxicity to humans and animals. This contaminant was isolated and identified in 1957 as TCDD, a form of dioxin. Although both 2,4-D and 2,4,5-T, purified of dioxin contamination, proved to be exceedingly poisonous, the extreme toxicity of dioxin even in immeasurably small doses attracted most attention.

By the late 1960s, controversy erupted over the safety of 2,4-D and 2,4,5-T. Reports accumulating from Vietnam, as well as from areas of the United States where phenoxy herbicides were used, implicated the chemicals as a likely cause of birth defects and other illness in animals and humans. Long-term damage to crops, trees, and whole ecosystems was also reported[10]. A study of the health effects of phenoxy herbicides by Bionetics Research Laboratories, begun in 1963 but not released until 1969, showed both 2,4-D and 2,4,5-T to be capable of causing birth defects at doses lower than lethal amounts[11]. These events awakened legislative and regulatory agencies to the importance of long-term, chronic exposure to chemicals in the environment.

In April 1970, prompted by Whiteside's articles in The New Yorker, hearings on the effects of 2,4,5-T on humans and the environment were held before the U.S. Senate Subcommittee on Energy, Natural Resources and the Environment, chaired by Senator Philip Hart[12]. These hearings led to the end, later that year, of the military use of Agent Orange in Vietnam. As for uses in the United States, however, the problem was not resolved by the hearings. Although both medical reports and the Bionetics laboratory studies showed 2,4-D to be at least as hazardous as 2,4,5-T, the spectacular toxicity of the dioxin in 2,4,5-T easily upstaged 2,4-D. On April 15, 1970, the secretaries of Agriculture, Interior, and Health, Education and Welfare announced suspension of the registration of 2,4,5-T for aquatic, home, and recreational use[13]. This was followed fifteen days later by a notice of cancellation of registration of 2,4,5-T for all uses on food crops intended for human consumption. (Suspension of a registration is effective immediately, without advance notice to the registrant or manufacturer; cancellation is effective only thirty days after serving notice, and is subject to the registrant's request for public hearing or referral to an advisory committee.)

Dow Chemical and other manufacturers of 2,4,5-T challenged the cancellation for use on rice, one of the major food crops treated with 2,4,5-T, and requested referral to an advisory committee. While the advisory committee deliberated, regulation of pesticides was transferred by Presidential order from the U.S. Department of Agriculture to the newly established Environmental Protection Agency (EPA). A year after the cancellation order was issued, the nine-member advisory committee submitted its report to EPA Administrator William Ruckelshaus, concluding that 2,4,5-T was unlikely to constitute a hazard to human health. One committee member, Dr. Theodore Sterling, dissented from the committee's conclusions in a powerful and eloquent minority opinion.

In spite of the advisory committee's majority opinion, Ruckelshaus issued an order three months later continuing the 2,4,5-T cancellation on food crops until completion of the public-hearing process. When Dow objected, the administrator issued a further order, reaffirming and supporting his decision with a ten-point statement of facts drawn largely from Sterling's minority opinion. Dow then successfully challenged the EPA decision in a U.S. District Court in eastern Arkansas (a rice-growing state).

"The curious thing to my mind," Dr. Sterling says, "is that to my knowledge of FIFRA [Federal Insecticide, Fungicide and Rodenticide Act], Ruckelshaus was supposed to have opened public hearings when he cancelled registration of 2,4,5-T. Now why would a shrewd lawyer like Ruckelshaus omit that? Subsequently, Dow was able to overturn the cancellation because he had not followed the procedures laid out by FIFRA. I suspected Ruckelshaus did that on purpose. He knew if he were to cancel 2,4,5-T, he would get the blessings of environmentalists, but at the same time he would get the blessings of the chemical industry because industry would force the courts to overturn his findings."[14]

After the Eighth Circuit Court overturned the district court's decision, EPA scheduled administrative proceedings against 2,4,5-T. These hearings were postponed repeatedly over the next eight years on grounds that more information was needed about the chemical.

The net result of this bewildering sequence of events was that during those years of regulatory jousting, Dow was able to market and promote a chemical in the United States that had been officially withdrawn from use by the military in Vietnam since 1970[15]. While Dow and the chemical industry successfully stalled EPA actions against 2,4,5-T, the controversy aired at the Senate hearings in 1970 had stirred legislators into action. In 1972, Congress passed major amendments[16] to the Federal Insecticide, Fungicide and Rodenticide Act intended to establish EPA's authority to ensure the safety of pesticides. FIFRA originally regulated only the labeling of chemicals. The new amendments required review and reregistration of all pesticides after EPA had determined their efficacy and safety[17].

The amendments, however, provided only illusory reassurance. Congress had passed the amendments, but Congressional oversight remained in House and Senate agriculture committees, which are notoriously responsive to the pro-chemical farm vote. Faced with over 35,000 registered pesticides to evaluate, EPA found itself pitifully ill-equipped, understaffed, and chronically short of funds to accomplish so gargantuan a task. Over the next few years,

EPA extended reregistration deadlines twice without completing reregistration of a single chemical. Meanwhile, pesticides continued to be sold under their old registrations, which amounted to false assurances of safety by EPA.

EPA took no further action against 2,4,5-T until 1979, when the agency ordered an emergency suspension of forestry use of 2,4,5-T, following a study of miscarriages in the area of our home (the Alsea study)[18]. Reregistration hearings on 2,4,5-T finally began in 1980[19], but were halted a year later when Dow and EPA entered secret negotiations to reach a "settlement" on the matter.

Although the hazardous nature of 2,4,5-T was well known long before 1970 when EPA was formed, the agency found itself bearing the burden of proof that by law should have rested on the manufacturers of the chemical. In addition, it found itself forced to devote its resources against a single herbicide and its contaminant dioxin, while other equally hazardous chemicals, such as 2,4-D, continued to be registered and used in far greater quantities[20].

By the mid-1970s the effects of Rachel Carson's warnings against insecticides were apparent. Some of the more lethal insecticides had been withdrawn from use, and the manufacture of others had declined significantly. As insecticides' share of the pesticide market declined, however, the herbicide industry prospered[21], its products generally escaping critical notice because their targets were plants, not animals. Although some insecticides had been banned or severely restricted in the years since Silent Spring, not one major herbicide had been so condemned. Industry was duly alarmed at the prospect of 2,4,5-T cancellation, openly expressing the "domino theory" that if 2,4,5-T should fall, other herbicides would soon follow[22].

Challenging even the least oppressive regulations, the chemical industry keeps regulatory agencies tied up in hearings and litigation, effectively concentrating agency resources and attention on one or a few chemicals at a time. In the case of health and safety regulations, a prolonged battle postpones questions of product or work-place liability. With chemical regulations in particular, except in the very rare event of an emergency suspension (exercised first against forestry use of 2,4,5-T in 1979)[23], a chemical may continue to be manufactured, sold, and used until the regulatory process is completed, which usually takes years. Industry protects its profits by prolonging that process as long as possible, challenging agency actions at every turn. Regulatory procedures established by Congress encourage such practices[24].

During the years that a chemical may be in the process of reevaluation—perhaps for grave health reasons—its manufacturer continues not only to sell it, but to promote it with safety claims already called into question by the regulatory agency. Four months after the EPA emergency suspension of 2,4,5-T in 1979, a "Dispute Resolution Conference" on 2,4,5-T sponsored by the American Farm Bureau Federation concluded that "...no known biological effects in connection with routine 2,4,5-T use have been documented over a 30-year period."[25] And six months later, as EPA prepared its case for reregistration hearings on 2,4,5-T, Dow Chemical wrote to a concerned stockholder, "There is no product that we manufacture that we have more toxicology and health data on than 2,4,5-T, and we consider it extremely safe."[26]

At the time both the above statements were made, Dow had already won an injunction against public release of the health and safety data it had submitted to EPA for reregistration[27]. Only close observers of the 20-year conflict over 2,4,5-T would know how unfounded such statements were. These industry claims exploit the profound difficulty of regulating poisons whose devastating effects may only surface after many years in the form of cancer, birth defects, or mutations, and whose immediate effects involve a broad range of symptoms (nausea, headaches, rashes, dizziness, psychological disorders, reproductive problems, weight loss, hemorrhaging) that are easily attributable to other causes[28]. It is virtually impossible to prove unequivocally a cause and effect relationship between exposure and symptoms, particularly when symptoms are delayed many years.

Unfortunately, chemical-industry interests have succeeded in appropriating a scientific tradition that requires unequivocal proof of cause and effect in its experimental methods. Laboratory studies, upon which science bases its conclusions, cannot possibly reproduce the conditions under which humans and animals are exposed to a chemical in actual practice. Common laboratory animals often differ significantly from humans in their sensitivity to particular chemicals. Because of ethical and legal constraints against scientific experimentation on humans, chemical regulations are based on laboratory studies which are often irrelevant to human or environmental conditions. Actual effects on human health in the field, because they are so difficult to prove or document, are thus dismissed as "anecdotal" or "circumstantial" evidence. Yet, while demanding unequivocal proof of harm, industry applies far less stringent standards to proving the safety of chemicals.

In the United States, it is illegal to perform pesticide experiments on human beings without their informed consent[29]. Laboratory rats are afforded no such legal protection. Rats cannot speak; people can. But since humans cannot be dropped alive into laboratory osterizers to have their tissues homogenized and analyzed, pesticide regulation relies on the mute testimony of rats and ignores the articulate voice of humans.

Scientists, regulatory agencies, and industry juggle lab animals and statistics while people like the residents of the Five Rivers and Alsea areas are exposed to phenoxy herbicides and suffer. Neither their exposure nor their suffering can be measured and charted. Their experiences qualify only as circumstantial (or "anecdotal") evidence in official judgments of herbicide safety.

Circumstantial evidence can never be conclusive, but it can be overwhelming. Only a human can testify to the effects of chemicals on humans. The parent is first to notice when something is wrong with the child. Only a forest dweller who knows the ways of the forest can testify to unusual changes there. Careful observation can never be replaced by controlled laboratory experiments.

No laboratory tests have ever "proved" that thalidomide or German measles virus cause birth defects in humans, for example[30], or that smoking causes lung cancer. Nor has it ever been proved scientifically that certain organisms cause particular diseases. Yet "the association is sometimes so obvious that you have to accept it," says microbiologist Patrick Schlievert. "If you ask anyone what causes typhoid fever, he will tell you that it is Salmonella typhi, but the causal relationship has never been proved."[31]

The pesticide industry, however, has successfully demanded scientific "proof" of human harm from a registered chemical before its registration can be withdrawn[32]. In 1978, Congress amended the pesticide law to provide that neither deficient laboratory tests nor "circumstantial" anecdotal evidence of human harm could justify cancellation proceedings against a pesticide[33].

After our children were hosed with 2,4,5-T in 1975, we learned that a rat in a laboratory cage commands more official respect than human first-hand experience. Under the law, the EPA can take action against a chemical only after proving unequivocal evidence of human harm.

Inadmissible Evidence

The Longyear farm, which had to be abandoned after helicopters sprayed the farm's watershed with herbicides, is about as remote from civilization as you can get in Oregon. It is accessible only by dirt and gravel roads which wind precariously through the coast range of the Siuslaw National Forest between Corvallis and the coast. No state highways or public campgrounds, no motels or roadside businesses tempt the tourist to the area. Travelers are daunted by broken lines representing dirt roads on the map. The timber industry, which controls most of the land here, has therefore operated largely unhindered by public scrutiny.

The countryside is a patchwork of living and dead vegetation. Occasional snags and massive, charred stumps are the only remnants of the vast old-growth timber that triggered the economic development of the Pacific Northwest. Where the road climbs through second-growth fir, you can begin to comprehend what it must have been like before, by imagining the trees as they might be a hundred years from now.

Up on the ridges, though, in all directions this second growth is being harvested, hundreds of acres of trees felled like soldiers in their prime. Nothing is left standing on these clearcuts. What is marketable is hauled out. All else is left to rot or burned to clear the land for replanting. On older units, "cat roads" up steep hillsides have eroded, carving great gullies and ravines. With no tree roots left to bind the soil, landslides have washed whole hillsides away, to silt up the river bottoms and destroy the spawning grounds of trout, steelhead, and salmon.

On more recent cuts, the land is littered with charred stumps, snags, unmarketable trees, slash and debris, all lying where they fell. These are visible

Image by Rio Davidson

Siletz River watershed, Oregon

Image by Rio Davidon

A clearcut and landslide (lower right) along the North Fork - Siletz River.

even from space, testimony to the universe of man's contempt for the greatest renewable resource the generous earth provides.

And yet, for all the devastation, the land is beautiful, with the frail, determined beauty of an organism struggling to recover from a critical wound or illness. Except for this year's cuts, the clearcuts miraculously are blanketed with green. The source of this miracle is the soil, and the weeds that feed it. Soil—mother of gardens, mother of forests—is itself the child of life, a living organism as complex and mysterious as a human brain. But it can support life only so long as it is fed by life.

Naked to wind and rain and sun, the soil cannot save itself.

Robbed of its wealth and life by man, it finds salvation in the enemies of man: weeds. Without weeds—all those plants which have no cash value— there would have been no forest here to cut. Weeds are the true and original pioneers of the forest. Tough, resilient, prolific and tenacious, they clung long ago to the crumbling rock of these ridges, bound it with their roots, drew water and minerals from it, thrived, reproduced, and decayed, each generation adding its remains to a newborn layer of soil. For how many centuries, how many millennia, did weeds work here to create sufficient layers of soil to grow a tree—or a whole forest?

Yet, in a matter of weeks, acres of trees are cut. The soil that weeds were building before man appeared on earth is laid bare, to be swept to sea by winter rains or parched and blown away in the summer sun. Left unhindered, however, the weeds return to save their own. Initially, the grasses, mosses, ferns, bracken, and a host of annuals move in, their airborne seeds and spores the first to find the waiting soil. In this new, tenuous vegetation, animals and birds discover a source of cover and food, and deposit seeds of food plants from other areas: salmonberry, blackberry, thimbleberry, salal, huckleberry.

Fast-growing berry vines and shrubs sprout and thrive in the shelter of the first arrivals, shielding in turn the seedlings of alder, maple, wild cherry, ash, and willow. Their roots hold the soil, and root bacteria replenish nitrogen. Their foliage retains water, shields the soil from sun and wind, feeds wildlife, and contributes a yearly blanket of leafy remains back to the earth. In this changing, varied forest of brush and deciduous trees, evergreen conifers (hemlock, fir, spruce, and cedar) will find a sheltered, nutrient-rich sanctuary for their primary growth, and eventually will outgrow and survive their deciduous foster parents.

Thus have weeds healed and renewed the land from the devastations of fire, flood, glacier, and pestilence since the beginnings of terrestrial life. Thus would it heal itself in time from the depredations of man, if given the chance. But out of the many species that support one another in a system as complex as a forest, man values only the trees that are immediately profitable. In the Northwest forest he sees only Douglas fir where once-prized stands of cedar, hemlock, and spruce were logged by his forebears. Alder and willow and other native species are weeds. Like the farmer who kills his chickens to collect the eggs, he mows down everything, strip-mining the forest to extract his profit.

The forester compounds the disaster of clear-cutting by attempting to replace instantly the forest created by generations of diverse plants. With explosives, napalm, fire, and poison, he bares the earth to replant marketable fir trees in weed-free stands, employing chemical warfare to suppress all else that grows. The forester looks upon his plantations of fir growing tidily like sugarcane, and is pleased with his work. He does not respect and cultivate the rich diversity that nourishes a forest. He is trained not to nurture, but to exterminate. From the arsenals of modern warfare, he has found a weapon that promises fulfillment of the exterminator's wildest dream: herbicides, the chemical defoliants developed during World War II to strip the countryside of vegetative cover, as well as to destroy vital food crops of the enemy.

Against these poisons, pioneer weeds have little chance to do their work of rebuilding a forest. They are killed off like dandelions in a lawn, replaced by species that are resistant to the chemicals but unpalatable or even poisonous to wildlife and livestock. One such species in the Northwest is tansy ragwort, a poisonous weed introduced from Europe that thrives on clearcuts, from which its airborne seeds rapidly colonize the surrounding countryside.

Chemicals used in forest management compound the reparable damage done by clearcutting. The results are readily apparent on the hillsides surrounding the Longyear farm. Recently cut units are strewn with slash and debris, wastelands of brown and black and gray tempered by the hopeful green of young ferns, bracken, grasses, and vines. Older units, left untouched since the timber was cut, are impenetrable jungles of salmonberry, salal, and vine maple. On the oldest units, young alders grow as thick as barley, a jungle paradise for deer and elk, countless birds, and small mammals.

But on the "managed" units, sprayed to clear them for replanting, the jungles are dead and brown. Dry skeletons of brush and alder and maple rattle in the spring sun. A towhee scratches forlornly in the brittle undergrowth. On a newly sprayed hillside, the foliage is still green, but pale and limp. The young shoots twist in grotesque, wilted spirals. An acrid smell of diesel and

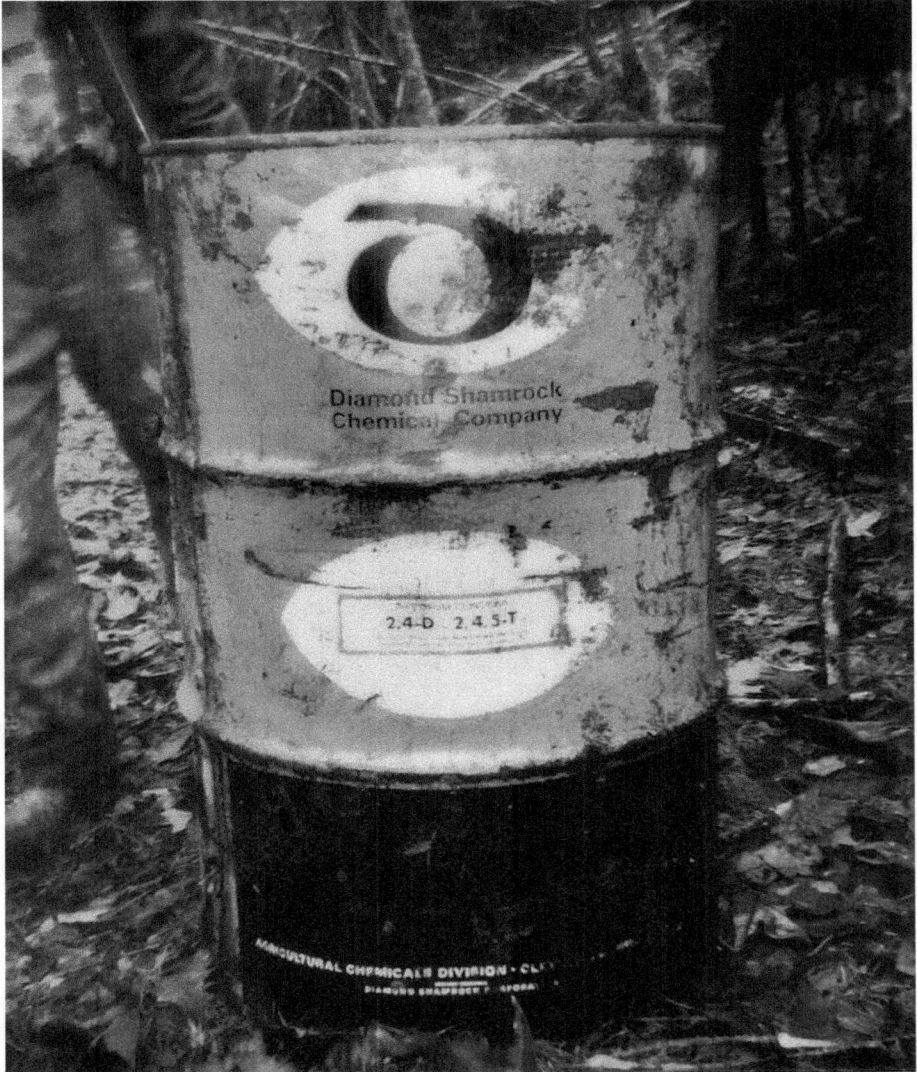

Image by Carol Van Strum

2,4-D/2,4,5-T (Agent Orange) barrel dumped on the Longyear's land

chlorine and decay vaporizes in the afternoon sun. Hundreds of yards away, blanched patches of adjacent formerly healthy forest testify to the effects of the poisonous drift.

The few stands of uncut timber welcome the eye. A yarder (a tower rigged with cables to haul cut logs to a landing to be loaded onto trucks) hums and whines in one such stand on the ridge above the Longyears', punctuated by the shrill bleeps of the chokesetters' signals.

The farm is an oasis of luxuriant green along the sloping valley of Rock Creek. Venerable, stout apple trees, green and in bloom, shimmer with clouds of bees and busy regiments of robins and goldfinches. Well-kept fences and gates border and divide the lush pastures. Strangely, no cattle or sheep or other livestock are visible. From a snag up on the ridge, two crows make nasty comments about the yarding rig. The place has an almost deserted quality, like a furnished house left abruptly by inhabitants who meant to return but never did.

Leon and Carrie Longyear moved here in 1944 with their young family to make a living off this land in any way they could. This was real wilderness country then. No power lines extended this far into the coast range, and the roads were no more than dirt tracks, often impassable. Leon and Carrie cleared land, planted apples, pears, nuts, vegetables, anything the land could produce. They raised cattle, dairy cows, and chickens for both meat and eggs. Everything was carried over those tortuous roads to the coast to sell.

It was not an easy life, nor, for a single family on a small remote farm, was it very profitable. After the years of depression and war, the security of owning land and producing their own food meant more than wealth or comfort. As the Longyears grew older and their surviving children grew up and married, power lines were brought in. The roads up to within a few miles of the farm were paved, others improved with gravel. By 1974, they were looking forward to an easier, more comfortable life in their eighties, when Starker Brothers, the owners of neighboring land to the east of their property, sprayed their watershed.

According to State Health Division records, Starker's land was sprayed on March 6, 1974, by Western Helicopter Company, with Dinoxol, an Amchem product containing 31.6 percent of 2,4-D and 30.3 percent of 2,4,5-T[1]. The Longyears remember that spraying. However, they are convinced that a second spraying took place, unreported, at a later date. They recall seeing the helicopter spraying over their spring sometime in April.

In April and May, the Longyears and one of their neighbors found five dead deer in a nearby pasture, and several more near their road. On April 28, Carrie became ill, with "paroxysmal vertigo and visual disturbances." Her condition worsened, and after collapsing in her kitchen she was admitted to Good Samaritan Hospital in Corvallis on May 1. She remained there for ten days, being tested and treated for "Organic circulatory problems in the central nervous system and hypertensive cardiovascular disease."

"We understand that water supplies on her property have been contaminated with 2,4-D and 2,4,5-T in the past," her physician, Dr. James A. Riley, reported in January, 1976. "The question is raised whether any of her present disease could be related to toxic levels of these substances in her body.

"My answer is that I cannot honestly say whether she has had toxicity from these substances, since no blood levels were taken or sought for at the time of her hospital admission or on February 17, 1975, because at those times we were unaware of the possibility of drug toxicity .... I could not definitely say that there was not drug toxicity, since these drugs can affect the central nervous system acutely, but as far as I know, their chronic effects are not known."[2]

Today, Leon and Carrie Longyear live in a small tract house in Corvallis, as close to Good Samaritan Hospital as they could locate. A "FOR SALE" sign is posted by the driveway to their farm, and the land they cleared and worked for 30 years awaits a new owner.

From the first, the Longyears' son Bob had been alarmed and angry about the spraying by Starker Brothers. He had taken it upon himself to collect water samples from their spring after the spraying. Putting together the dead deer right after the 1974 spraying, Carrie's illness, and their son's misgivings, the Longyears began to suspect the herbicides sprayed by Starker as a common cause of their troubles. They filed a complaint with the Oregon Department of Health, and wrote to Senator Bob Packwood of their problem.

State Health Division records state that Starker personnel had drawn samples of water from the Longyears' spring on March 6, 7, and 8, and sent them for testing to the Environmental Health Sciences Center at Oregon State University (OSU). The results of these tests were that no residues of the herbicides were detectable at a detection limit of less than two parts per billion[3]. (A detection limit is the smallest amount of a chemical able to be detected and identified with certainty by the particular method of testing used.)

Apparently influenced by the OSU test results, the Health Department's investigation concluded that "Mrs. Longyear's illness and the dead deer were not associated with herbicide application on Starker Forest, Inc., land." James Kirk, the investigator for the Health Department, supported his conclusions with five rather questionable observations[4]:

1. Absence of detectable residues in water samples taken after the application. [At the time, Kirk had only the OSU findings to go on, so his conclusions from this testing are understandable.]

2. Long interval (53 days) between application and the onset of Mrs. Longyear's symptoms. [This does not take into account the rate of stream flow, heavy precipitation subsequent to spraying, or the Longyears' claim—evidently never investigated—that the land and spring were sprayed again during April.]

3. Medical diagnosis of Mrs. Longyear's illness reveals nothing related to pesticide poisoning. [This was based on a telephone message relayed to Kirk from a physician not referred to in Mrs. Longyear's hospital discharge, with no actual discussion of the case with an attending physician. Evidently neither Kirk nor the doctor he consulted made any attempt to research possible symptoms of pesticide poisoning or to relate such research to Carrie's case. Her attending physician, Dr. Riley, noted later that the herbicides "can affect the central nervous system acutely," and that "I could not definitely say that there was not drug [herbicide] toxicity."]

4. The treated area presently supports a substantial deer population. [This observation was made on October 1, seven months after the

spraying, and has little relevance to the finding of dead deer within days of the spraying. It is difficult to understand how these live deer could prove conclusively that dead deer found seven months earlier "Were not associated with herbicide application..."]

5. No unusual fish mortality at the hatchery [a State Fish Commission hatchery that raises coho salmon downstream from the Longyears' stream]. [Kirk's report does not state that the hatchery is located fully five miles downstream from the Longyears' spring, nor does it estimate how many other streams feed into Rock Creek in that five mile stretch, thus diluting any pesticide present. He reviewed hatchery records for mortality only, not for hatching, developmental, deformity, or other problems.]

What appears to be a thorough investigative report by the Health Department turns out to be based on observations (of trees and deer) made seven months after the event in question. It belittles Mrs. Longyear's symptoms, noting suggestively that her family did not relate her illness to the herbicide spraying until after her hospitalization. Kirk concluded his report by accusing the Longyears of requisitioning several discarded herbicide drums dumped by the spray crew at the helicopter landing site, including a partially full one. He neglected to point out that such dumping was prohibited by the herbicide label. Furthermore, he either neglected or deliberately ignored the fact that the label on the discarded drums was not the product reported by Starker Brothers.

Dismayed by official response to their complaint, the Longyears decided to have the samples their son had collected from their water supply tested independently, but had difficulty finding a lab they trusted in Oregon. In July, 1975, they finally sent the samples to Morse Laboratories, Inc., in Sacramento, California. The Morse lab found 20.9 parts per million of 2,4-D and 4.03 ppm of 2,4,5-T in the water[5], levels which certainly should have been apparent at OSU's lower detection limit of 2 parts per billion.

Carrie sent the results of the Morse lab tests to Senator Packwood, who referred them to Edwin L. Johnson, Deputy Assistant Administrator for Pesticide Programs of the Environmental Protection Agency. Mr. Johnson's response reviewed the flimsy status of 2,4,5-T in EPA's estimation at the time. Johnson included a memo from the EPA's Dioxin Monitoring Program reporting findings of TCDD, the deadly contaminant of the herbicide, in beef from rangeland treated with 2,4,5-T and expressing concern about the implications of these findings to human health[6]. Johnson outlined EPA's current and prospective 2,4,5-T monitoring plans, dismissed any suspicion of 2,4-D as a threat to health, and regretted that EPA had no responsibility beyond "The area of environmental pesticide control."[7]

This informative letter from Mr. Johnson was interesting to the Longyears, but promised them little help and no comfort in their present situation. By this time, both Carrie's and Leon's health had deteriorated to the point of almost total debilitation. Carrie was unable to do more than putter about the house, and Leon had developed cancer of the colon which required surgery. He also acquired an acute, recurring diabetic condition that led to amputation of his left leg.

"It didn't make no sense by then," Carrie says, "And it still don't. Them EPA people send us all these papers, but I never could see the meaning of it, except that they knew already what bad stuff it was, and they knowed it was in our water, and they just wasn't going to ever do a thing about it. So there wasn't anything more we could do, and we was too sick by then to do it if there was. We couldn't run the farm anymore; the stock was all sold 'cause none of 'em could breed no more, and we couldn't neither of us live so far from the hospital by then. So we give up, I guess you could say. We give up and put all the papers in a box and moved away."

The Longyears' son Bob still lives in a small trailer house with his own family on the farm where he grew up. Since a recent heart attack, he has been unable to work, and stays on to keep the place up as best he can for potential buyers. A sturdy man in his mid-forties, Bob Longyear is someone who would seem at home on a logging crew or combine. Instead he putters restlessly about the trailer and garden, or his pickup truck, which he is working on now.

"No, it won't be easy selling all this. I don't know what we'll do. Move to town, I guess. I was here by myself and had this heart attack—hit me sitting right here in the pickup. There was no one else around, and I just fell on the wheel and waited it out. I had to drive up to Washington that week, and that pain was like nothing you can imagine. It just didn't let up. I went ahead to Washington anyways, taking aspirin every couple hours to keep the pain down.

"I knew it was a heart attack, but you know you never want to believe it's that serious, and so I went along, taking all that aspirin, and it corroded my gut, you know what I mean? Everything got ate away and started bleeding, and that's what put me in the hospital. I just got back last week. Doctors couldn't understand it, neither. I just lost 34 pounds, and they couldn't understand it, losing all that weight and then having the heart attack, and up to then I was always so healthy."

Bob tugs at his belt to show the gathered waistline of his pants. While he talks, a chainsaw whines on the ridge. A yarder bleeps. He starts involuntarily at the familiar sound, then shrugs.

"There's just no way Mom and Dad could've stayed here. I don't know when Dad got diabetic. It just got bad these last few years, and then he lost his leg from it. It got the gangrene in, and I don't know what all they done to try to save it, but didn't work. And Mom couldn't work like before. She just got too tired since that time she collapsed. She just went down, you know, like a horse.

"That was in '74. I found her in the kitchen, right there in the doorway. I took her to the hospital myself, and you know what the first thing the doctor said was? Right there in the emergency room, he checks her over and first thing he says, 'Has she been exposed to some kind of spray?'"

"Of course we said no. We was thinking of fly spray around the house, Raid or one of those, or something we might've sprayed on the fruit trees, you know, and we hadn't been using anything like that. We was thinking of something she would've got sprayed right on her, or handled, you see. We didn't think at the time of it being in the water, of her drinking it for weeks.

"Starker told us they was going to spray that ridge, in March. I told them 'No way: That creek is our water supply, it comes from a spring up there.' I told 'em O.K., only if they stayed 200 yards from the spring. That was fair, wasn't it? But I could tell they wasn't going to pay me no mind. I got the state forestry people up here and asked them to stop them.

"So they come up here with their notebooks—them forestry people— and I show them the situation with the creek and our water supply, and they stand here and tell me the stuff—the spray—was safe. So I ask them flat out if they would line up here and drink some of that stuff, and not one of 'em would. Not one. But they wouldn't do nothin' to stop it."

Bob pauses and leans on the tailgate of the pickup. "There wasn't nobody else you could go to in those days. You try to get help anywheres, and they all tell you the same thing: that it's all in the hands of them forestry people, the agriculture boys. Now I ask you, whose side do you think they're on? And it ain't like I got anything against forestry, against logging. Lord knows, we made a livin' off the timber from this place—that's what paid all my folks' hospital bills! That's a good joke, ain't it?

"So Starker goes ahead and sprays. And they sprayed right over the spring. We saw 'em do it. And afterwards the water was downright milky here. Scoop it up and it was like a bucket o' milk, with a kind of oily surface, you know, and you could smell the diesel in it. They mix the stuff with diesel.

"And let me tell you something else. They—Starker, that is—can swear up and down a stack of Bibles they took water samples, but that just goes to show how much respect they got for anything, even the Bible, 'cause if they took samples it was from somewheres else. They sure as hell didn't take any water from this creek. They couldn't have without my seeing 'em, and I was waitin' for 'em. We took our own samples, and you know about that. We sent some of the water to OSU.

"Ha! As if you can trust OSU any more than the damn government. They're all the same—like General MacArthur—they'll say anything you want to hear, and go ahead and do as they damn well please. And you know who founded OSU, don't you? Well, T.J. Starker, who else? He was one of the founders, and he was head of forestry there for thirty years[8]. So you know how much you can trust that outfit." He stares at a redtail hawk, directly over-head. Over the crest of the ridge, the sound of the chainsaw is broken by the staccato cracking of timber and the dull thud of a tree hitting the ground.

"So they sprayed. They said they only sprayed once, in the beginning of March, but I know better. We saw 'em up there again in April, saw the helicopter, right over the spring again, too. And the containers my neighbor found up there in late May was too fresh to have laid there since three months.

"And that's another thing. Them labels. The report Starker give the forestry boys and the health department, they say they sprayed with a Amchem spray, Dinoxol, it says. But look at them labels," he says, pointing to a herbicide drum at the side of the trailer. "That's a different spray, Weedone. Enough to make you wonder, ain't it? But we didn't notice that then, and nothing happened right away, you know.

"The water cleared up, and we was all right, so when Mom collapsed we just weren't thinking of the spray up there. When the doctor asked about sprays we was only thinking of something that might've happened that day, or the day before. At a time like that, you aren't thinking of things real clear, you know. You don't stop to think how that stuff takes time, how it gets in your system and builds up there, and by the time you get sick from it, it's too late to do anything about it.

"I wasn't thinking of all that when I told 'em to keep the stuff away from the water, though. I wasn't really thinking of us, of it hurting people. I just knew it was bad. I knew that a long time. Back in the forties, the extension people did a demonstration out there in Toledo. That was before they moved the courthouse to Newport, it was in Toledo then. And they took a whole bucket of that stuff—2,4-D and 2,4,5-T mixed – and sloshed it over a patch of blackberries there next to the courthouse, takin' pictures and all to show how wonderful the stuff was, you know. And a few days later, when I saw how the path where they slung the stuff was killed dead, I knew.

"You know how hard it is to kill blackberries? That's how I knew. Anything that could kill a blackberry, I knew it couldn't be no damn good. And I'll tell you something else. They knew it, too. In '76 they sprayed again. And when I saw how this water looked, I took one look at it, all milky with that oily scum on it, and I called the Fish Commission down there at the hatchery— that's five miles downstream from here—and I told 'em they better get up here and test this water...They come here and take a look, and tell us the water looks

O.K. to them. They said it was safe to drink. Just from looking at it, they tell us that. And you know within 24 hours they had moved every damn fish out of that hatchery. That's at a time of year I never knew them to move fish before—'cause those fish are worth money, see, but what are *we* worth?"

A young girl comes out of the trailer, carrying a bucket.

She is about eight years old, still dressed in her school clothes.

"If you're watering the chickens, you got the wrong bucket," Bob tells her. "You got to take the big one. And pick 'em some greens, too. They got to have greens, now they're shut up."

He retrieves a crescent wrench from the grass and leans on the tailgate, opening and closing its jaws. "But I knew long before what this stuff could do. Most folks don't know. They don't even connect what's wrong with 'em with the spraying. I got a friend who was an applicator for years, spraying the stuff. He still don't know, but you go see him, you take one look at him, and you know. He's dying, and he don't know. It affects the brain, honest to God, you don't make connections any more. Maybe you don't *want* to know, 'cause by the time you're sick you'd know there was nothin' you could do.

"Even Mom. After she come home from the hospital, she come out one day to help move one of them containers they left—my neighbor bring 'em down from the helicopter pad where they dumped 'em—and you know she didn't even touch one of 'em. Just walking near the thing her arms broke out so bad and she got so dizzy she had to go back inside. That's where she realized how bad the stuff was, how sensitive she was to it.

"In the fall of '76—that's *five months* after they sprayed that year—I was hunting with a couple of guys up on another unit where they sprayed along the road. We was elk hunting, and it started to rain, and you could just smell the stuff steaming up in the rain along the road. It made me so sick I had to beg off and come home. First time I ever quit an elk hunt.

"They say it's all coincidence, what's happened to us. But look at what's happened—*all since 1974*. There's my dad. He got this cancer and had to have a hunk of his intestine and what is it—his colon—took out in surgery. And then he gets the diabetic thing and has to have his leg cut off. He don't believe it's the spray, 'cause he wouldn't drink this water after they sprayed. He took his water from another spring. That don't signify, in my opinion, 'cause you know how that stuff gets in the air. It don't ever stay where they put it.

"But maybe it wasn't the spray. If that was all that happened, just my dad's troubles, I wouldn't think nothing of it. But then Mom collapses, and you know after she come home from the hospital she didn't use this water neither, not for months. Then in February she thinks it must be 0. K. by now, and she starts drinking it again, and what happens? She collapses again, and she ain't been right since.

"And you got to understand, it ain't like they was sickly city folks or nothin'. These are folks used to workin' hard and stayin' healthy in this kind of life.

"And I'll tell you something else," Bob continues thoughtfully. "I haven't seen a live fawn around here since 1974. Not one. Used to see them all the time. I grew up here, you know, and I know these woods and animals. You grow up in this life, you grow up as tough as the next one, you know, and I never thought twice about hunting and all when the time come. But there's some things you just come to look forward to, if you know what I mean, and I never missed seeing the fawns. But there hasn't been a one since 1974.

"And then there's the cows. After that year not one of 'em had a live birth. Not a single one. We sold 'em all off, finally. If you don't get calves you don't keep up your herd.

"And only one chick—just one single one—has hatched since 1974 on this place. Put the eggs to set, and nothing happens. You crack 'em open finally and there's nothing, just rotten. And no more geese hatched since then, neither. And then I lose all this weight the doctors can't understand, and have this heart attack. You couldn't prove any of it, but it all adds up, don't it?

"They say it's all a coincidence, but that's a few too many coincidences for me to swallow."

Early Warning

Maybe all of you are just going to sit in Washington, with your doors and your windows and your minds closed, and hope these chemicals don't drift your way.

—Billee Shoecraft[1]

Overcutting of timber is not the only way to abuse land. In the western United States, large areas of public rangeland have been destroyed by overgrazing of livestock. When too many animals are grazed on too little land for too long, they consume the edible grasses and other forage plants faster than the plants can replenish themselves. Eventually, the only surviving vegetation consists of unpalatable or poisonous "weeds" which spread rapidly in the absence of competing plants.

In the arid mountains of Arizona, both overgrazing and timber cutting have taken their toll on the landscape. In 1965, the U.S. Forest Service began an herbicide program in the Tonto National Forest to reduce chaparral (thorny scrub oak, manzanita, and other shrubs that thrive on overgrazed areas).

"The ultimate objective in the vegetative change is to obtain and maintain a more or less open semi-desert type with desirable shrub and trees for game cover and open grasslands in between... to increase water yields and forage for domestic livestock," the Tonto Forest supervisor said[2].

The Salt River (hydroelectric) Project invested in the Forest Service program. In theory, clearing the watersheds of vegetation would increase water runoff for Salt River Project reservoirs. In August, 1965, a Forest Service ranger wrote in a column in the Globe, Arizona, local newspaper:

> One of these days soon, you'll look up toward the old C.C.C. [Civilian Conservation Corp] site and see a helicopter flying repeated trips over the brush country near the head of Six Shooter Canyon, Kellner Canyon and Russell Gulch. The helicopter will be dropping a herbicide spray.
>
> If you're as curious as I am, you'll want to drive up and watch the operation. I hope you will.[3]

From 1965 to 1969, the USDA Forest Service sprayed thousands of acres in the Pinal Mountains near Globe with various mixtures of 2,4-D, 2,4,5-T, and 2,4,5-TP (silvex), the same chemicals being sprayed over the forests of South Vietnam at the time. Except for three or four large-scale ranchers, no residents of the sprayed areas were notified, either before or after the spraying. In the first years of the operation, the few residents who saw the helicopters and planes at work—or who complained of ill effects—were assured by the Forest Service that the spray was perfectly safe, harmless to all animal and human life.

Unless a resident connected problems with the spraying, the Forest Service did not suggest such a connection. In 1966, Mrs. Billee Shoecraft asked a ranger what was causing the pine trees to die on the side of the mountain overlooking her property. She did not know at the time that the area had been sprayed. The ranger told her only that the trees were afflicted with a "mysterious disease."[4]

Over the years of spraying, the McKusick family, who lived at the end of Kellner Canyon, had complained of herbicides drifting onto their land whenever neighboring canyons were sprayed, particularly after they were sprayed by an airplane on their own property in 1968[5].

One morning in June 1969, Charmion McKusick heard a helicopter making passes over their home. She ran outside into a choking fog of herbicide swirling and dripping on her children playing in the yard, on the goats, chickens, trees, and stock pond, and on the laundry hanging from the line. The McKusicks' lives have never been the same since.

"You talk about silent spring," said Charmion, a former avian specialist for the National Park Service. "It was still. It was eerie. Everything was dead—not a mouse or a ground squirrel survived."

In the months after the spraying, the McKusicks found blind, undersized calves, deformed horned toads, ducks with bill and wing defects, a goose born with its vital organs on the outside of its body, a paralyzed peacock, chickens with crippled legs and droopy, oversized wings, and young goats and cats with misshapen spines and heads, sometimes one eye above the other. Their cats' eyes and ears bled for the rest of their lives.

"Not no way can you tell me this is normal. We lived here fifteen years and never saw this in animals, and all of a sudden 70 percent are deformed and wiped out," Bob McKusick said. "But they [the Forest Service] have an answer for everything. They don't give a damn." The McKusicks were told at various times that the damage was caused by fire, road oil, grazing horses, woodpeckers, and insects.

The McKusicks' soil is still contaminated eleven years later.

"I still can't work the soil with bare hands without breaking out in blisters," Charmion said. Her husband gets back and chest pains, stiff neck, headaches, weak legs, and a skin rash when he comes in contact with the soil. Charmion has had a precancerous lymph gland removed, intestinal surgery, and a hysterectomy, and still coughs blood and has vaginal bleeding. Bob has had several heart attacks since the spraying, attributed to ventricular fibrillation. The cartilage between his joints is dissolving, making him shorter.

"We can see by our animals that we're still having problems," Charmion said in August 1980. They had just shot a pet goat with a tumor the size of Charmion's fist on the side of its head. "We thought we were going to get better, but we're going to die. We're used up. We're not good for anything. We're going to die. There's no amount of money on the face of the earth to give us back what we've lost."

The McKusicks were not the only residents sprayed directly by the Forest Service helicopter that June morning. Bob and Rosalie McCray, at work on a house they were building along Kellner Canyon Road, were preparing lunch for their five-month-old son when the helicopter flew over with its spray nozzles wide open, soaking them, the baby, and Nacho Hernandez, who was helping them build their house. Hernandez became sick within minutes of the spraying and was taken to the hospital. He later developed oozing sores over much of his body, similar to those suffered by workers at factories manufacturing phenoxy herbicides. For the next four years, the McCrays' baby, Paul, was not expected to live. He was having up to thirteen seizures a day. Doctors could find no explanation or treatment.

An anonymous phone caller who identified himself as a government employee saved Paul's life. He instructed the McCrays to tell their physicians that the herbicides were causing anemia, a red-cell suppression and a runaway white-cell count, inducing convulsions in the child. Paul's doctors acted on this advice, and he survived. In 1980, at the age of eleven, he was doing well, although he had a cyst from membrane deterioration that required surgery.

The McCrays' goats produced no healthy offspring during the first year after the spraying. Four kids died at birth or shortly after. One that survived (after its mother died giving birth) had a malformed jaw. The poultry started laying eggs without shells and eventually died off. Rosalie McCray still has health problems. She discharges milk at odd times, a complaint shared by Charmion McKusick. She feels run-down, and her daughters have never had normal menstrual cycles.

Bob McCray had already been exposed to herbicides in 1968, at the service station where he was foreman. One of his employees had had a seizure while steamcleaning herbicide drums brought in for cleaning by the Forest Service, and Bob had finished the job. His eyes became bloodshot and light sensitive for days, and he had chest pains and a "tingling sensation" in his hands and feet. The worker who had the seizure died several months later. Three years ago, Bob had surgery to remove cancerous connective tissue (fibrosarcoma) in his shoulder.

From a campsite in Kellner Canyon that same June morning in 1969, three sixteen-year-old boys started on a hike up into the mountains and were suddenly engulfed in a cloud of herbicide. One of the boys, Jim Beavers, remembers "An ugly smell, overpowering, like weed killer." His head started pounding and waves of sickness made him too dizzy to walk. He collapsed onto a picnic bench and was taken home by a deputy sheriff. The pounding headache continued for six hours, and his heart began to have "flutterings." He thought he was going to die. Shortly thereafter, he dropped out of high school.

"I wanted to do some things before I croaked ...That's what was going through my mind." A few years later, Jim Beavers and Mark Nowell, one of the other boys on that hike, became fathers of children born with Down's syndrome (mongolism).

In yet another home on Kellner Canyon that morning, the Steinke family were eating Sunday breakfast with guests when they heard the helicopter coming in low over the house. "We ran outside. It [the helicopter] was spraying a fine, wet mist, and more chemicals were leaking out the bottom," Wilma Steinke said. Her daughter and son-in-law jumped into their truck and followed the helicopter to its landing area. "The spray was drifting everywhere," Wilma said. "My daughter came back covered with it. They were all excited because it was so new to us."

Within three days, Wilma and both her daughters were hemorrhaging vaginally. "My mom thought my insides were falling out. There were big clots, tissue, like meat, coming out," Betty, the younger daughter, said. She was fifteen at the time. Her older sister, Jean, who had followed the helicopter with her husband, had a miscarriage five months later, followed by severe hemorrhaging and coma from loss of blood. The hemorrhaging continued for three months. A specialist who had experience treating women exposed to Agent Orange in Vietnam was called in by hospital physicians. It was "highly probable," he said, that Jean's problems were related to herbicide exposure.

The Steinkes' cow had difficulty calving. Another calf had to be butchered after it developed breathing problems and couldn't suck. It was found to have an enlarged heart and liver and a shriveled lung. The family's fruit trees withered, bloomed out of season, and died. Their shade trees died. Eighteen laying hens died within the first three days after the spraying.

Forest Service and USDA officials refused to acknowledge that any problems along the canyon were associated with herbicides. The supervisor of the spray project later claimed that the big mistake the Forest Service made was in spraying on Sunday when people were at home. USDA investigators attributed the damage in the canyon variously to insects, disease, woodpeckers, sapsuckers, low soil moisture, and air pollution from a nearby copper smelter.

"They insult the intelligence of many people," Wilma Steinke said. "They don't think you notice what happens day in and day out."

The water in the Steinkes' well tested at 5 parts per million 2,4,5-T, and the Arizona Department of Health found 3.4 parts per billion silvex in the municipal water supply at Globe. Yet the only illness USDA conceded might be related to the spraying was "eye irritation and skin rash" suffered by the service-station worker, who collapsed while cleaning the herbicide drums and later died.

Probably the biggest mistake the Forest Service ever made in spraying Globe, Arizona, however, was the spraying one Sunday morning of Billee Shoecraft. It may be that newcomers to an area become more fiercely devoted to their land than old-time residents, perhaps because they are there by choice rather than necessity. Certainly no one was ever more fiercely devoted to a piece of land than Billee Shoecraft was to the twenty acres she and her husband Willard bought in the Pinal Mountains.

On that Sunday morning of June 8, 1969, the U.S. Forest Service helicopter sprayed Billee Shoecraft as she stood in front of the house she had built, drenching her and the pink chiffon nightgown she was wearing. "The result was astounding," Dr. Frank E. Egler wrote:

> One hundred pounds of love and adoration for her family, her pets, her garden, her home, her mountains—finding all this jeopardized – was metamorphosed into such an energized force that I even believe an irate mother grizzly bear would be put to shame. In this case the force was compounded with intelligence, self-education, direction, persistence and faith. It is the sort of syndrome that lesser mortals bow and bend before, and try to preserve their own egos by calling it "emotionalism" and "paranoia."
>
> [Her] book is the story of what happened after that fateful nightgown-wetting day in June, 1969, up to the time that she slapped a $4.5 million lawsuit against the U.S. Forest Service and four herbicide manufacturers. It is a story of laughs and tears; of incredible bureaucratic bungling at county, state and federal levels and within industrial circles; of disillusioned dreams and of anger and contempt, of whitewashing,

cover-ups, alibis, legality, duplicity, arrogance, conceit and crass ignorance; and of a few knights in shining armor.[6]

Billee Shoecraft's crusade against the poisoning of her world quickly evolved into a crusade against the corruption, indifference, and ignorance she encountered. Her book, *Sue the Bastards!*, marked the birth of a powerful, growing anti-herbicide movement that Billee herself did not live to see.

In the weeks following the June 1969 spraying, as foliage and animals died and her family sickened, many unexplained incidents from the previous four years became clear to Billee. She began checking dates, and found that the strange "sickness" of the pine trees on her mountain correlated with a 1966 spraying. Deaths and birth defects in her son's guinea pigs correlated with other sprayings. Her notes during the three months following the 1969 spraying included: Three dead doves; bird dead by back door; bird dead in pool. Two dead skunks, dead fox, on hillside. Dead bobcat by bridge by house; dead bugs on ground; four scorpions in living room; three centipedes in lumber pile. Many "children of the earth" bugs in pond. Dead fish in pool; long pink centipede in house. Dead bobcat by mailbox and two male deer by patio wall. Dogs did not bark at them; baby guinea pigs keep dying.[7]

At the Globe radio station owned by Billee's husband Willard, callers to the "open line" program reported similar occurrences. Deteriorating health plagued them all. Willard Shoecraft, whose legs had been amputated as a child, could not wear his artificial limbs, his discomfort was so severe. The Shoecrafts' son had back pains, nose bleeds, and eye problems, and developed a limp. Billee herself suffered from nausea, retching, a throat disorder that obstructed speech and swallowing. Her arms swelled, with open, dripping sores that defied treatment, an ailment she remembered from other dates when spraying had been done.

One thousand people from this small mountain area signed a petition sent to the Tonto National Forest supervisor. Copies were sent to state representatives, to Senators Barry Goldwater and Paul Fannin, and to the chairman of the U. S. Senate Subcommittee on Environmental Resources. By the middle of July, only a few noncommittal replies had resulted. The petitioners' request for a stop to the herbicide program was ignored by the Forest Service. A petition to President Nixon with several thousand signatures also failed to stop the Forest Service from continuing its program.

"A lot of illusions were shattered after the spray of 1969," Billee wrote, "And a lot of them were mine."

When the petitions failed to bring any response or concern, Billee ran up a $500 phone bill in one month, trying to find a government official who would listen. She made calls to state agencies, Forest Service officials, the Army Corps of Engineers, the Bureau of Indian Affairs, the Food and Drug Administration, and the Department of Health, Education and Welfare in Washington, D.C. She called the Bureau of Land Management, USDA, and "some of those big doctors at Beltsville." She called the adjutant general, the President's science adviser, and even President Nixon himself.

"We called them all. And what happened? Nobody gave a damn!"

Perhaps if someone had given a damn when a small community of Americans appealed to their own government for help, the issue would have died quietly. If the U.S. Forest Service had acknowledged it might have oversprayed or allowed drift to contaminate private land, or officially declared its spray project finished, the residents of Globe, Arizona, would have considered their petition answered. Their shock, disillusionment, contempt, and anger at their government's dismissal of them were in direct proportion to their shattered faith in that government.

Official neglect of the situation in Globe was one thing. Official lies were another. If a government agency announces it has put men on the moon, or denies it has bombed a neutral country in Southeast Asia, a citizen can either accept its word on faith or cynically disbelieve. But if your own yard is full of dead trees, birds, and animals that an official task force reports to be "hale and hearty," mistrust becomes not cynical but eminently reasonable. Billee Shoecraft lost faith in her government in 1969, but she never lost faith in herself. Neither did she lose faith in American democracy. Indignant and outraged at its degeneration, she set out to correct it.

Undaunted by official titles, military brass, top-secret rubber stamps, or agency protocols, Billee embarked on a singular campaign to find out what was behind the lies, secrecy, and indifference Globe residents had encountered. Billee once defined herself as "poet, designer, architect, writer, mother, and friend...most of all, an American, and a citizen of the world." She was not trained in science or politics, but she knew how to read and she knew how to learn. Tirelessly—through phone calls, trips to Washington, endless foraging in files, following up every clue and reference—she amassed in a few months more information on the herbicides sprayed in Globe than any agency or official knew existed.

Armed only with the title of American citizen, Billee badgered, cajoled, and intimidated officials in almost every branch of the U.S. government, as well as scientists under contract to them. She scrutinized government regulations, court records, scientific studies, congressional reports, and military documents with a fresh perspective. Some of the material she unearthed had been withheld from the public, including the series of studies performed on the phenoxy herbicides by Bionetics Laboratories for the National Cancer Institute, mentioned in Chapter 2. The studies indicating that both 2,4-D and 2,4,5-T were capable of causing birth defects in small doses had never been released, even to other government agencies.

Billee's information met little response from the government, particularly the Forest Service, the state of Arizona, or the Salt River Project. Numerous "Task farces," as she called them, arrived in Globe from government agencies and research units. Most left without seeing more of the sprayed area than could be seen through an official car window, and their conclusions were that no evidence of herbicide drift or damage could be found. Scientists who visited the area were unable or unwilling to respect the testimony of "unaccredited" residents. Politicians avoided any involvement in a program so dear to USDA, the chemical industry, the military, and a major hydroelectric company.

One scientist and one politician finally did hear the pleas for help from Globe, Arizona. They not only listened, they came and looked, and acted upon what they found. In February 1970, Democratic representative Richard

D. McCarthy of New York visited Globe with Arthur Galston, a biologist from Yale who may have unwittingly contributed to the development of phenoxy herbicides with his early research on plant growth hormones[8]. A few weeks later, they reported on their investigation of the Globe incident to the hearings on phenoxy herbicides of the Senate Subcommittee on Energy, Natural Resources and the Environment, chaired by Senator Hart[9].

The disastrous effects of Agent Orange in Vietnam, the haunting specter of *Silent Spring*, and *Life* magazine photos of limbless thalidomide children reinforced the testimony on herbicides that Globe residents presented to Senator Hart's subcommittee. In addition to halting military use of Agent Orange in Vietnam, the hearings led to the 1972 amendments to the nation's pesticide law, requiring that all pesticides currently on the market be reregistered under new testing procedures that included testing for mutagenicity, carcinogenicity, teratogenicity, chronic effects, sublethal effects, and effects on nontarget organisms. The amendments set up even more rigorous testing requirements for new chemicals not yet on the market, including the provision that they be tested before registration. Under the new provisions, a pesticide must be re-evaluated and reregistered every five years. Responsibility for pesticide regulation was transferred from USDA to EPA, and on paper at least, EPA was given strong regulatory and enforcement authority.

In 1970, however, even Senator Hart could not move the USDA. The residents of Globe abandoned hope of governmental assistance or action. Late in 1970, the Shoecrafts, McKusicks, and four other families filed suit against the Salt River Project, four chemical companies, and a helicopter company. They were unable to sue the government for damages until other legal remedies had been exhausted[10].

In liability cases of this nature—as in the notorious legal suits against the distributor of thalidomide in this country—industry attempts to settle out of court, in effect buying the silence of the plaintiff, who can seldom afford a protracted court case. But the Globe families were determined to hold out for a court hearing. Unlike an out-of-court settlement, a court decision in their favor would establish a precedent for liability in other cases. For eleven years, as the defendants won repeated postponements, the Globe families held out. (By 1980, only Dow Chemical remained of the original six defendants named in the case. The claim against the Salt River Project had lapsed. Several other defendants settled, but the families said they never received any money.)[11]

"Dow is trying its very best to keep any of the cases from coming to trial," said Dorothy Thompson, an attorney involved in a nationwide lawsuit initiated in 1979 on behalf of 4,000 Agent Orange veterans. "Attrition is one way to win a lawsuit. You wait for the plaintiffs to die off. You wait for the evidence to become so stale that the other party has no way of marshaling its forces."[12]

At least one of the plaintiffs has died off. Billee Shoecraft died in 1976 of a cancer that had been discovered, noted, but allegedly not reported to her by a physician examining her at request of Dow Chemical Company, several years earlier[13]. Bob McKusick estimates that nearly twenty people in the sprayed area of Globe have died from illnesses associated with the herbicides[14].

In March 1981, the day of the opening of the trial, the weary Globe families settled out of court for undisclosed terms. As part of the settlement, Dow admitted no liability[15].

"We'll never get old, just die suddenly one of these days!" Billee wrote in 1971. "I had my tombstone all worked out before this spray. It was to read, 'She had a ball and didn't miss anything.' Now I've had to change it to, 'See, I told you I was sick!'"

Drifting in the Wind

Indian Creek is a tributary of the Siuslaw River in the southern portion of the Siuslaw National Forest. This is a productive area of the Siuslaw, still actively logged. Its valleys support fruit growing, hay crops, cattle pastures, and truck gardens, with easy access to markets both in Eugene and in Florence.

The Wilkinson farm is in the hills above Indian Creek. Characteristic of long-standing small farms in western Oregon, it was created not for profit but for self-reliance. It produces food for a family which is dependent on the husband's job with the timber industry. Pastures, garden, and orchard have been carved from the forest, supporting a few steers, a dairy cow, occasionally a pair of weaner pigs, a flock of chickens, a pony or horse for the children, and perhaps a goat or two to keep the blackberries in check.

The Wilkinsons' house is functional and unpretentious. It is small, as places reliant on wood heat tend to be, added onto and altered by the family as the need arose. Outside, rose bushes, lilacs, and daffodils flourish in the spring rain. Elk-head trophies from successful hunts dominate the interior, staring benignly at the worn, comfortable furniture, the framed photographs of children, the television screen.

Of all occupations, logging is a way of life closest to that of the early pioneers in this country. The logger with a family typically lives on a farm such as the Wilkinsons'. He is paid well for grueling, dangerous work, and is prepared to be laid off without notice at any moment. He and his family survive at the whim of weather conditions, mechanical foibles, and the economic vicissitudes of the timber industry. They grow most of their own vegetables and fruit, raise their own meat, eggs, and dairy products, preserve their supplies by canning and freezing, and supplement their meat supply by hunting and fishing.

Their food and livelihood come from the forest. So also does their recreation: hunting, fishing, camping, hiking. In remote country such as this, electric power and telephone are unreliable amenities. Self-reliance and a strong, traditional family structure are matters of survival. In the long absences of her husband, a housewife and mother finds herself taking on the chores of carpenter, plumber, farmer, stockman, or mechanic as the needs arise. Her children grow up in a world where survival requires quick wits, toughness, and willingness to work hard and together. Television, mass culture with its self-indulgent values, and the busing of children long hours to distant centralized schools take their toll on many logging families, particularly recent settlers whose roots are elsewhere. But the Wilkinsons cling to a way of life handed to them by several generations of both their families in the same coastal forest.

"My husband was born here," Jackie Wilkinson says.[1] "His folks live right up the hill there. His mom raised five kids in a cabin with only a dirt floor. I was born in Yachats [on the coast, due west over the hills], so we've both lived in this area all our lives.

"We've lived in this house seventeen years. In the early years, it was a whole day's trip to go to town. There were no paved roads then – it was just logging roads between here and my folks' place. It's easier now, a lot of things are easier, but our life hasn't really changed all that much. We still grow our garden, and keep a cow or two and the chickens, and Harry gets his elk, and a deer or two, and goes out for the steelhead."

A kettle whistles from the kitchen. Jackie moves to its command, making coffee, setting cups, cream, and sugar on the table.

"We have the television, and that's nice. Sometimes there's something good on it, a good program or two, and I like the news. Harry will come home from work and lose himself in the T.V. He doesn't even care what's on it, it just helps him relax. But except when I can't sleep, I don't watch it all that much. The kids used to watch it more, especially my daughter, but Andy was mostly too busy—fishing, hunting, sports, all kinds of projects, and drawing. He loved to draw, and he was really good. He did that painting of the elk up there, and these cards."

The painting is of a bull elk, standing in the dappled light of a forest clearing. Trees loom over it on both sides, the tree trunks behind it lost in a haze of green. It is not so much a picture of an elk as a glimpse of forest and elk as a single being.

Jackie spreads on the table a collection of greeting cards and sketches, proudly pointing out her favorite, a grinning "raccoon cartoon."

"He liked drawing animals best. Andy died in February of this year. He'd been sick for two years—not just sickness and pain from the cancer, but from all the treatments, the chemotherapy and radiation they tried to cure it with. I don't think anybody but me could know what he went through those two years."

She leaves the drawings on the table, glancing at them often as she talks in her calm, soft voice. "It started in May of '77. He hurt his knee in track – nothing broken or anything, just wrenched it going over a hurdle. But it never seemed to heal, and the pain kept coming back. Finally in February of '78 the pain came back and was so bad we took him to the doctor. He had an x-ray taken, but it didn't show anything.

"Then in March the pain got so severe there was no question there was something very wrong. Our doctor sent him to a specialist, who took another x-ray—only six weeks after the first one. There was a big change in that time. The bone looked motheaten, the doctor said. He suspected some kind of infection, and so Andy went back for surgery.

"Even then, the doctor thought it was an infection, but when he did this culture—you know, growing some of the bone cells—it showed a cancer. They didn't know what kind it was at first. From the culture, it seemed to be something they called a non-Hodgkins lymphoma.

"So we took him to Portland for more surgery. By that time he had a sore spot on his head, too. They had trouble typing the cancer there, too, and they sent a culture to the Mayo clinic. They agreed with the first diagnosis, non-Hodgkins lymphoma of the thigh and hip.

"They started chemotherapy and radiation treatments. I drove him to Eugene—that's eighty miles each way—every day for those, and to Portland for checkups...He would try to joke with the doctors, to make it easier for them.

He knew all along, of course. The doctors never hid anything from him. They asked us at the beginning if they should tell him, and we said of course, it was his problem, his suffering. And he knew, anyway. On that first trip to Portland, our pastor came with us—for support, you know—and Andy told him that day, 'I know what it is. They can't do anything, I know.'"

She sits back on the couch, sipping her coffee. "He had so much spirit. No one ever wanted to live as much as he did. It was frustrating for him, he tried so hard and was in such pain all the time. After the first year of treatments it seemed like it might be working. He seemed almost well. And then in July of '79 he had a relapse. The cancer had gone into the bone marrow. It had become leukemia, and didn't respond to any treatment after that ... He died in February in 1980. He was thirteen...

"We asked the doctors once if they knew what had caused the cancer, and they said they didn't know. We had seen about the sprays, and the trouble they cause, and we asked if it could have been that, and they said no. But if they didn't know what caused it, how could they be so sure what didn't cause it?

"We live here, so far from all kinds of pollution, we've always been healthy, our food and water come from the forest, and there's no history of cancer in either of our families. The sprays—the herbicides—are the only unusual thing that has come into our life here, the only kind of pollution, I mean. Of course we can't prove anything, but if the sprays did cause the cancer, I wouldn't want any parent to have to go through what we did with Andy—the sickness and the pain and the weakness, knowing how it was going to end, knowing that he knew. If he had been hit by a truck or something it would be easier. It would have been sudden and over with, and we wouldn't have his suffering to remember. If there's even a chance the sprays caused it – well, I wouldn't want to hear about anyone else going through what we did.

"You can't live out here without getting sprayed, one way or the other. In April of 1969, we got sprayed directly, even. By the Forest Service. We had taken the dogs out hunting rabbits, and had come up Green Creek from Deadwood. We parked the pickup right alongside the road, where they had clearcut a few years before. It was good hunting there where the brush and alders had come up. Andy was with us – he was only two then – and his sister. She was four, two years older than him. We were right there in the open, walking up the hill from the pickup.

"There was no way the helicopter pilot couldn't have seen us. He came up over the hill, spraying. It was a long time ago, and I couldn't swear to it, but I believe he came back a second time and sprayed right over us again.

"We didn't know anything about the sprays at that time. We didn't even know what he was spraying, but we were very upset and angry. We weren't thinking of ourselves at all, of it making us sick or anything. It was the thoughtlessness of it. It was rude. There was no way he couldn't have seen us. And one of the dogs we had then, we knew it would make her sick, because any kind of spray made her sick. She'd get so bad she couldn't breathe. We were thinking more of her than of ourselves.

"When we talked to the Forest Service, that's what we were thinking about, and about their carelessness. And it wasn't until several years later that we talked to them at all, after one of our neighbors had trouble with them spraying around her place. They—the Forest Service—came out and taped an interview with us, but at the time we didn't know anything about what kinds of symptoms the sprays caused, and I don't think we could report any sickness or other effects from that particular incident.

"I think they wanted us to say for sure that we hadn't been affected, but they don't think of the effects that could happen years after, do they? I don't remember all we said then. The Forest Service only this year finally told us what it was they sprayed us with—a mixture of 2,4-D and 2,4,5-T, a pound of each per acre. They have never answered my request for the tape. We were bashful at the time about making a big thing out of it. It wasn't important to us except that it made us so angry, but one thing I do remember about that interview was my husband said if it ever happened again he'd shoot them out of the air.

"So I don't think we had anything to tell them in that interview about health effects, not any that happened immediately, anyway. But that year, the year we got sprayed, something strange did happen to me. I got some kind of nerve problem. That's what the doctors said it was, anyway. I was tired all the time, my joints ached, I had skin troubles, and I couldn't handle things at all, not even the children. They had to go stay with my mother. The doctors said it was just nerves, but the medications they gave me just made it all worse, and they had to put me in Sacred Heart Hospital for five days. They did all kinds of tests, but they couldn't determine the cause. It wasn't anything I'd ever had before. We didn't know anything about the sprays at the time, and of course we didn't make any connection, but when you look back you begin to put things together.

"They spray all over around here. It's part of living here. If you don't know anything about the herbicides, you don't think of noticing when or what they're spraying. You go through the sprayed areas every time you drive anywhere, and of course they spray all the roadsides, too. That road to Yachats we take to go to my folks' place, that must be sprayed every year. You can smell it. And everything—all the plants and little trees—are dead. Except the tansy [tansy ragwort, a noxious weed]. The tansy just comes back worse than ever.

"You know, my folks used to pull all the tansy on that road, all the way over to here, every time they came over, but now there's so much you couldn't touch it. And I've noticed whenever they spray the roadside you don't see the little animals along the road any more. The little diggers [ground squirrels] and rabbits and chipmunks—they just vanish when they spray.

"When they spray the forests—with the helicopters—you can smell it here even when they're a distance away. One year, they were spraying down the hill there, and the helicopter would come up this far to turn around. You could see the spray coming out as he turned, right over that field. We didn't get any fruit at all that year, but we really didn't think about it. They just have a job to do and you believe them when they say this is the best way to do it. They're the experts, aren't they?

"We're always told to trust them. They're the experts. They know.

"But then things happen. Not just one thing but all kinds of things, and then you get to thinking about it, and you begin to wonder. They say it's safe because they fed it to some mice and they didn't die. But how long does a mouse live? Two years, maybe? How can that prove that a child, say, won't be sick from it in five or ten or even twenty years? It would make you nervous about nursing a baby around here. But then the cows eat the grass, so it's in their milk, too, so what can you do? It gets everywhere.

"Look at the St. Helens volcano. It blew up last night at—when, about 10:30? By this morning we had ash all over our windshield, and that mountain is 150 miles away! So you know that the stuff they spray five or ten or more miles away, that's going to land here, too. You just can't see it like you can the ash. Something like a volcano—well, there's enough things you can't do anything about, but something like the sprays, that's not a natural thing."

In the field outside, a small sorrel horse is trotting about in the rain, tossing her head and leaning over the gate to look up the road. Jackie stands by the window, watching her.

"She's waiting for the school bus. You know, besides losing our fruit, we've had strange problems with our animals over the years. Things older folks around here don't remember from the old days. There was the dog, and we knew about that. She would get sick just from walking on the roadside when they sprayed. But the cows we didn't know. They were calves, six or eight of them, all ages from six weeks to six months old, all very healthy. Usually what you lose calves to is scours—you know, they get diarrhea, you're always watching out for that.

"But this was different. They were beautiful, healthy animals, and suddenly their whole back end would go out of control. They wouldn't be able to stand, and in two days they were dead. We had the vet autopsy them, and he sent them to OSU for testing, but they couldn't find anything.

"They said it was probably tansy. That's what they always say when an animal dies and they don't know why, they say it's tansy. But I've seen an animal that's sick from tansy. It works on their liver, and they get real poor. They're sick for a long while and just get worse until they die. That's not the way it was with these calves. They were fine and healthy, and in two days they were dead.

"Our neighbors brought a wild horse home and the same thing happened. They had him six months or so and he was just fine, and then his back end went and he died the same way the calves did...

"It makes you wonder when they say tansy is killing all this stock. I know it's poisonous and I've seen animals die of it, but not that many. The cows just plain don't like it. They'll pick over the hay and eat all the grass and leave even the small leaves of tansy. I don't think they'd eat it unless they were starving and there wasn't a twig of anything else. But now whenever any stock dies, you send it to OSU for tests and they say it was tansy. So it comes out that all these thousands of head are being lost each year to tansy and they scare people into spraying it real heavy.

"One of our neighbors up the hill lost a big Charolais bull, the same way our calves went. They told him that it was tansy, too. But he doesn't have any tansy up there. He keeps it down by hand spraying all the time. Isn't that enough to make you wonder, though?

"After the calves died, and no one could tell for sure what happened to them, I asked the vet was it possible they could have been poisoned by something, and he looked at me like I was some kind of paranoid nut-case, like I was suggesting some enemy had sneaked up in the night and given them a shot of something with a hypodermic needle. The doctors were the same way when I asked about the sprays causing Andy's cancer.

"They look at you like you're crazy, but the truth is, they don't know either. They don't know any more than I do about what caused Andy's cancer, or the calves dying, or my sickness, and they don't know either what those sprays can do in a person's system over the years."

Outside, the little mare races to the end of her field at the sound of the school bus approaching. Moments later, Jackie's daughter and a friend come in, their hair wet from the rain. They dump their school books, rummage in the kitchen for a snack, and are off into the rain again to the friend's house.

"It's kind of scary, you know," Jackie says. "She's only two years older than our son. If it was the spray—if it was his age at the time that made him susceptible—then could it affect her at a later time, or maybe in some entirely different way?

"Our kids were always real healthy, except for the usual colds and flu once in a while, but now she's got that hanging over her. It's hard enough for her, because she and Andy were very close. When you live out here there aren't many other kids close by, and they only had each other. Now he's gone, and she has to live with this big question mark over her.

"It isn't fair. We're grown, we've done a lot, we've had a life, but the kids haven't ...

"I'd give anything I have not to have this happen to someone else. But they're still spraying."

* * *

The Wilkinsons' neighbors, Jean Anderson and Ugo Pezzi, live on their 1300-acre ranch along Indian Creek, where they moved in 1967 to raise prize Charolais cattle for the organic beef market. Jean is a clinical psychologist in private practice in Eugene, and Ugo is a geologist. In 1971, the Forest Service sprayed several units bordering Jean and Ugo's land. The herbicide drifted over their pastures, where their cows—some of them due to calve soon—were grazing. Over the years of Forest Service spraying in the neighborhood, Jean and Ugo noted many problems with their livestock: stillbirths, stunted growth, sterility and reproductive problems in the cattle, and fatal crippling due to nervous-system damage in a quarterhorse mare. Not only was their certification to sell organic beef threatened by the Forest Service spraying, but the proposed spraying program for 1973 would also endanger the watershed for their domestic water supply.

After several years of ineffective pleading with Forest Service officials and lengthy scientific critiques of their environmental impact statements, Jean and Ugo requested a public hearing in 1973 to discuss their research on herbicides and their concerns about the effects of the spraying. The Siuslaw Forest supervisor denied the hearing, and they proceeded to file suit in federal court, asking for an injunction against the spraying of herbicides in their watershed. In the end, they agreed to withdraw the suit "without prejudice" in return for an agreement by the Forest Service not to spray in their watershed. ("Without prejudice" gave them the right to reopen the case at any time if the Forest Service ever violated the agreement.)

"The spraying was a serious threat to our whole operation here," Ugo says[2]. "There are a total of 193 streams feeding this watershed. There was no way they could spray without contaminating them. But the real issue to me was that of the basic human rights involved. I happen to be very strict about my right to privacy. I don't care if they were spraying rose petals on our land and water—if they do it without my knowledge or consent, they are violating a basic right to privacy."

"It goes further than the question of invasion of privacy, really," Jean adds[3]. "It's a question of voluntary or involuntary exposure. One of their favorite arguments is that herbicides are no more toxic than table salt or aspirin. Aside from the fact that table salt and aspirin do not accumulate in food chains, taking them is a voluntary action. People can choose whether to take them or not. But the person who is exposed to the chemical droppings of the U. S. Forest Service has no more choice or control in the matter than the peasants of Vietnam did."

* * *

When Salty Green and Gisele, his beautiful West German wife, bought an old homestead in 1972 in the Alsea valley, their ultimate goal was to grow organic meat and produce in sufficient quantities that Salty could quit his salesman's job and devote full time to farm and family.

Over the years, they have transformed the original 100-year-old settler's cabin into a lovely, comfortable home, with all the fine attention to detail that an old-world heritage often loses in America. The house and all ninety acres of the farm—pastures, barns, flower and vegetable gardens, orchard, ornamental plants, field crops, and forest—reflect that heritage, the creation of something beautiful and enduring from what the earth provides.

The Greens' two boys, ages ten and eight, are bilingual, articulate, and intelligent, equally at home baking in the kitchen or helping with haying and other farm activities. The extent and diversity of those activities are extraordinary. On their 65 acres of logged-over land, the Greens have replanted 6,000 trees, maintained with a nurse crop of clover to provide food for deer, competition with weeds, and a reservoir of soil moisture. On the remaining acreage, they raise steers for organic beef, maintain twenty acres of irrigated pasture, grow vegetables for home and market, tend a sizeable orchard of fruit and nut trees and a small commercial berry operation, and raise sheep, geese, ducks, chickens, rabbits, and pigs.

After four years of hard work, the Greens were ready in 1976 to begin marketing certified organic meat and produce. Salty's goal of financial independence seemed imminent. But on April 3, 1976, the Greens were wakened at 7 A.M. by the sound of tank trucks driving up their road into the timberland adjacent to their farm, part of the large area around Alsea controlled by BLM (Bureau of Land Management). An hour and a half later, they heard the sound of a helicopter spraying the forest.

Four years later, Salty is still absent from the farm for months at a time, unable to quit his salesman's job. Trees around the house and garden still show signs of damage from BLM herbicides that contaminated their property in 1976 and prevented the organic certification they had worked so long to acquire. Gisele manages their diversified farm alone during Salty's long absences, still bewildered and distressed about the spraying four years ago, and the behavior of the government agency responsible for it.

"Salty followed the tank trucks up to the helicopter landing when they first came. The BLM man in charge up there was very nice. He said they had checked the wind and there was no problem. He told Salty they wouldn't be spraying within 100 feet on each side of the creek and that they planned to stay good and clear of our property. But by the time Salty got back, it was blowing strongly here—the tops of that tall cedar tree were moving, and my hair was blowing in the wind. It was obviously blowing more than six miles an hour, and the helicopter was spraying.

"At a meeting with the Forest Service a few weeks before, they had told us about regulations that prohibited spraying when the wind was greater than six miles an hour. I called the state forestry office in Philomath and the Forest Service in Alsea, but they said they had no jurisdiction over BLM While I was on the phone, the helicopter came right over our place. They were spraying all around us and took several turns above my greenhouse. Some friends arrived just then, and while we were all standing on the porch the helicopter started taking its turns right over my garden. I called everybody inside because I didn't want any spray to get on the children and dogs.

"I fixed breakfast, but I was very upset. Around ten o'clock we were all watching because the helicopter was coming very close again, making his

Image by Sam Shin

A helicopter sprays herbicides on a clearcut in Oregon

turns right over our field. The pilot would swoop down over the trees, across the road, and right out over the middle of our field. The sprayers were on full blast, and even when he turned it off the spray was like water dripping out of the back of the nozzles. You could see how the wind was blowing the spray all over. All the animals were running around. The chickens and cows were just terrified.

"Salty went back up to the helicopter landing, quite angry. He told them they were spraying our fields, that the spray was blowing all over, and showed them on the map where we own property on the other side of Honey Grove Road. They said they'd be more careful, but even after Salty got back the helicopter came in close again while I was milking. I was hysterical. It was so very awful to see the helicopter coming over like that, to see that spray coming down on our field, on the garden, on everything, and not be able to do anything to stop it. And I was afraid for the animals. They were hysterical, too."[4]

The helicopter did more than spread terror among the Greens and their animals. It destroyed four years of hard work in a matter of hours. To sell meat or produce with organic certification, no chemicals may be applied to the land for three years. After four years of preparation, Gisele had arranged in 1976 to sell organic produce from their farm. The helicopter set her marketing plans back another three years.

"That night, you couldn't walk out the door, it smelled so bad," Gisele says. "Not a diesel smell—it was definitely a poison smell. You could still smell it a good week afterward. Two days later, the BLM spray supervisor came out. He said as far as he knew, the pilot had followed his instructions to stay away from our land, but of course he couldn't see down here from the landing, and when we pressed him, the most he could say was that the helicopter had taken off in the right direction.

"After that, we started finding oil slicks—from the diesel they mix the spray with—in puddles and water troughs all over our barnyard and pasture. I called BLM and they sent people out to take pictures and collect samples. The oil slicks were all over the pastures and way up into our forest, more than 800 feet from the BLM land they were supposed to be spraying. We walked all around our neighbors' places, too, and found the diesel spots there as well."

The Greens' friend and neighbor Dr. Daniel Elam accompanied Gisele and the BLM representatives on the inspection of the oil slicks. Dr. Elam, a retired doctor of chemistry who worked in the petrochemical business for 25 years, lives at the foot of Honey Grove Road along the bank of the Alsea River. In his small garden, orchard, and ornamental plantings, he experiments with organic methods of horticulture. Retirement has freed him to pursue his scientific interests more actively than ever. What was happening on the Greens' farm and his own land piqued his scientific curiosity and led him to some valuable observations about drift patterns of aerially sprayed chemicals.

Dr. Elam carefully marked in red the location of each diesel spot on an aerial photograph of the Greens' property and the surrounding area, his red marks showing clearly how far from the target area the herbicide traveled.

"Two points must be made in respect to these oil spots," he says. "It is not important to note the quantity of the spots in any location. What the spots show is how far the drift traveled. The second point is that these oil droplets fell all over the region where we found them. It's only in the puddles of water that they could be seen."[5]

Within a month, the Greens noticed severe damage to their trees, crops, and pasture plants. Gisele took samples of the withered, curled leaves to OSU and to the state agriculture department. Dr. James Witt at OSU was certain the leaves had been damaged by herbicides, but Tom Harrison of the Agriculture Department declared that the damage was due to aphids and worms. Nor was Gisele able to find out what chemicals had been sprayed on their land. BLM claimed they had no record of what was sprayed, and Evergreen Helicopter Company would not give any information.

"It became very obvious that neither BLM nor Evergreen was going to admit responsibility for spraying our land," she says, "Although the vegetation samples taken to Dr. Witt showed residues of 2,4-D, we never were able to find out what had been sprayed, and everyone seemed to have a different story. The helicopter pilot told us at one time that 2,4,5-T was sprayed. BLM told us that it was only 2,4-D, and told OSU to test our samples only for that. But when the state Department of Agriculture ran tests, their report listed the chemical used as a mixture of 2,4-D and silvex.

"As spring advanced and the leaves reached maturity, it became obvious that most deciduous trees on our place had been affected in some way or other. Most of them showed stunted growth, and had only half the leaves they normally would have had. The tall ash at the end of the driveway showed half-size leaves which were curled under around the whole leaf. One half of the elderberry tree alongside the barn died—the same as occurred to trees in the direct-spray area. The pear tree looked very poor, and although the grapes and mulberry tree still grew, they looked very sad. Their fruit was mostly dwarfed, with only a few normal size grapes.

"That was all four years ago. We took Evergreen Helicopter and BLM to court. They never would admit they had done any damage here, and probably never would have, except that the helicopter pilot finally, in his deposition, stated that he had sprayed across our boundary. Even after he admitted that, they wouldn't admit responsibility, but offered to settle with us out of court.

And we accepted. I know we could have carried it further and made a bigger case out of it, but it had already taken up a whole year of our lives, and we just wanted to be rid of the whole affair.

"At that time, there was just no support anywhere for our position. Everyone seemed to think we were either just nuts, or that we were just out to make a lot of money in court. But that wasn't it at all. It was the principle of the thing – that they should be able to come over our land and do so much damage and we couldn't do a thing to stop them, and then they could get away without even admitting what they had done.

Lambs born on Green farm, 1980[6]	
Jan. 12	triples, one ewe lamb deformed
Jan. 13	Winnie aborted
Jan. 14	Bummie aborted
Jan. 19	Blimpy's lamb born dead, deformed front legs
Jan. 24	black ewe's lamb born dead, deformed legs
Feb. 23	ewe #90 red tag, severely deformed lamb (malformed spine, legs, and tongue), died Feb. 25, lived 36 hours
Mar. 26	Black yearling ewe aborted a lamb
Three aborted, One deformed, lived Two born dead, One deformed, died	
Out of 13 ewes that lambed, 6 lambs died, one deformed lamb lived, 9 normal lambs lived. Seven out of 13 ewes had problems with their lambs.	

"The damage wasn't just for that year. It destroyed our chances for years to come of selling our meat and produce on the organic market. If you look out there now, you can see that our trees have never recovered. They still have the same symptoms they had four years ago. You wonder if that stuff just gets in their system and stays there somehow. Or does it just change their growth patterns permanently? No one has the answers. But it changes things for good, it seems. And other disturbing things have happened. Our cow aborted the next year after they sprayed. And this year there has been a frightening development," she said. She displayed a chart showing the problems with her livestock that year.

"Here is a record of the lambs born here on our place in January. Out of fifteen births, seven lambs were deformed, and six died. I just tell you, we never had anything like this happen before. And right under that is a record of lambs born in February to ewes at the Stouts' ranch a mile past Honey Grove Road, right over the ridge from us. It's discouraging, isn't it? My first thought was just to sell all the sheep, take them to the auction, it's too much to worry over. But how could I pass on such problems to someone else and have that on my conscience? And then, too, I felt curious, I guess, to see what happens next year. But it's frightening, too. Could there have been something that's been building up in those ewes for four years since they sprayed?"

Lambs born on Floyd and Mimi Stout farm, Feb. 7-9, 1980. (Farm located one mile south of Honey Grove Road, Alsea)		
Feb. 7	Twins	#1 died at birth, deformed chest, curled front & back legs
		#2 died at birth, deformed chest, curled front & back legs
Feb. 8	Triplets	#1 Ram, OK
		#2 Ram, live, now OK, knuckled over front posterns at birth
		#3 Ram undersized, died at birth
Feb. 9	Twins	#1 died at birth, no deformity noted, but wouldn't breathe
		#2 died after 24 hours, curled front & back legs, deformed chest
Feb. 9	Single	Died at birth, curled front & back legs, deformed chest & mouth
One ewe died and autopsied, found 3 lambs all with curled legs, deformed chest and mouth. Same ewe had a single lamb last year with curled front legs that died at birth. Lambs were born with heart beating, but didn't breathe.		

"Our own health, and the children's, has not suffered, but I do not think that has great meaning, because I have strong belief in the power of the mind over the health of the body, and we have refused to think that the sprays could cause us any sickness. We have always been very healthy. "But the spraying has affected our whole life, since it happened. We were just farmers, trying to grow healthy, organic food in a beautiful place. This whole thing has made me aware how little some people care, how they abuse the environment, and how they proceed to cover up what they've done so they can go ahead and do it again. I could have gotten more active, more involved, made more public an issue of it, but there's only so much you can do, because you still have a life to lead, and a family. That's what comes first – my family – and I could only fight this up to the point where it would interfere with that."

* * *

At about the same time the Greens were noticing the damage to their trees and crops a month after the spraying, Dr. Elam had become puzzled by the sudden withering and dying of the vetch, grapes, and apple leaves in his own garden. "Could some kind of plague have hit us?" he wondered. He had taken some samples to the county extension agent, who could not explain the cause. The OSU plant clinic, however, immediately diagnosed the problem as phenoxy-spray damage.

"But that spraying was two miles away!" Dr. Elam recalls. "I must say, I couldn't quite believe then that it could have reached us and affected all our plants so strongly."

In June, Dr. Elam began noticing the vine maples turning red, "Bright red-the way they do in autumn." Strange clumps of yellow leaves appeared on the big leaf maples near his home. He drove up to the Greens' and noticed the same "blemish" on the trees up there.

"The OSU plant clinic could find no pathogen responsible and attributed the death to a 'phytotoxic' agent. I began to explore the area systematically then, to see if there was any sort of pattern in the occurrence of this malady. I found that the yellow patches were present all the way from the upper part of Honey Grove Road down to the highway, and on the hillside across the highway opposite Honey Grove Road. East on the highway they occurred as far as the Stouts' place, and as far west just to the border of Alsea, which is about three quarters of a mile from here.

"Looking at a map, I put together an explanation. When they sprayed there above the Greens', the day ended in a warm night, with an incomplete cloud cover. Now you have to understand what happens when that stuff is sprayed. No matter how accurately they spray, it's impossible to control what happens to it afterward. Evaporation will always occur. Now the herbicide is a heavy molecule. It vaporizes, but in the cool of the evening it settles close to the ground and then proceeds to flow along the ground.

"What evidently happened in this event was that it flowed across the Greens' place down into the whole Honey Grove valley. Here, at the lower end of the valley, it ran up against the hillside across the highway, then spread laterally along the highway as far east as Stouts' and west to the edge of Alsea.

"As you can see on the map, or just by looking at the terrain here, this valley served as a kind of sink for the spray. You could certainly smell it here that night after they sprayed. It was fierce. And apparently it was strong enough to damage the vetch and deciduous trees and grapes. Who knows how long it went on draining down here? For as long as it continued to vaporize off the sprayed area, one can assume.

"One thing that really makes me wonder—the deformed lambs and stillbirths in the sheep both at Greens' and at Stouts'. And Mimi Stout had a dog-breeding operation, and many of their pups were born deformed recently, too. They won't put any blame on the sprays, because they use them themselves. They say they've used them for years and years and never had any problem. That's a common argument of the spray operators, but what they don't realize is that for most of those years the spraying was done by hand and was applied only locally. There wasn't the possibility for widespread drift and accumulation that aerial spraying causes. Aerial spraying by helicopter is relatively new, and I don't think these people realize what a greater potential there is for drift and damage."

The continuing damage to the Greens' trees and sheep is a visible reminder of uncertainties that have never been resolved about herbicides. The public interest that has grown steadily since the Greens were sprayed in 1976 was almost nonexistent in Alsea. Salty and Gisele were essentially alone in their battle, and the knowledge of law, bureaucratic procedures, and toxicology developed by citizens' groups since then was unavailable to them at that time.

In the social climate of 1976, to sue BLM and the spray contractor was an extraordinary move. Few families even now have the resources and temerity to sue an agency of the federal government on their own, and fewer still can afford to endure years of legal haggling by refusing to settle out of court. Gisele's involvement did not end with that settlement, however. In 1980, she and Dr. Elam flew to Washington, D.C. to testify in behalf of EPA, telling of their experiences with drift damage at cancellation hearings on 2,4,5-T and silvex.

"I have great faith that EPA will act to protect us from these occurrences in the future," Dr. Elam says. "In every contact with them, working with them on these 2,4,5-T cancellation hearings, I have been impressed by what dedicated, hard-working, conscientious people they are. You know, Dow has tried to promote the image that the case against herbicides was mounted by a bunch of hippies, dope growers, and radicals. I think EPA has not been fooled. They have made an effort to meet the people out here, and become acquainted with the kinds of people concerned, and I think they have a pretty clear picture of what's going on. They are on our side all the way, and I have been happy to work with them."

Not Only Plants

In recent years, a pleasant, grandmotherly woman has been called to testify at various hearings and criminal trials on the hazardous nature of the herbicide 2,4-D. Her name is Ruth Shearer, a molecular geneticist and consultant in genetic toxicology from Issaquah, Washington. She is often called to testify in support of damage claims filed by people exposed to toxic chemicals, speaks at public meetings and hearings on the subject of phenoxy herbicides, and has been called as a defense witness to support a citizen's action in halting a spray program near his home.

Dr. Shearer's appearances at such occasions are due not to advocacy of any particular cause, but to the fact that she is one of the very few scientists with expertise in the genetic effects of 2,4-D and its relationship to cancer. After the emergency suspension of 2,4,5-T in 1979, 2,4-D was often used as a substitute for it. Human health problems continued to occur after exposure to 2,4-D, and concerned citizens found that compared to the massive research devoted to 2,4,5 T for many years, very little research was available on 2,4-D.

"I'd been doing basic cancer research for ten years before I was asked by Metro, a public agency in the Seattle metropolitan area, to do a world-wide literature search on four herbicides, including 2,4-D." Dr. Shearer says. "I did this in early 1979. The report was published by Metro in January 1980. It was a study of the health effects of the herbicides on mammals, and studies that involved human exposure and injury."[1]

Dr. Shearer, visiting friends in coastal Oregon, stands at the window of their home, holding their two-week-old baby. The baby, nestled comfortably in her arms, stares intently at her as she talks.

"I found that in the whole world's literature, only three cancer tests have been done on 2,4-D, none of them adequate by today's methodology. Two of them were done in the U.S. They used too few rats and mice, and the mouse experiment used too low a dose and too short a time for an adequate cancer test. However, both of these tests did give some statistically significant positive results.

"The third test was done in the Soviet Union. They used more than adequate numbers of mice and rats. As an assay for complete carcinogenesis—the ability to induce all stages of cancer—this test was negative. But they also ran a test for cancer promotion, to see whether 2,4-D would stimulate premalignant cells—that had already been initiated by another carcinogen – into fully malignant cancer cells. This test was very strongly positive.

"Now 2,4-D contains as a contaminant another chemical, 2,4-dichlorophenol. Also, 2,4-D is broken down into 2,4- dichlorophenol by microbial action in the environment. The 2,4-dichlorophenol is an even stronger cancer promoter than 2,4-D, and is a weak carcinogen—cancer initiator—as well."

The baby, his eyes closed, holds one of Dr. Shearer's fingers in his small fist. He smiles fleetingly in his sleep, and she smiles back.

"2,4-D induces mutations in both animal and human cells in culture, and damages DNA in a manner similar to ionizing radiation—x-rays or gamma rays." Dr. Shearer continues. "It causes developmental toxicity in offspring when given to the pregnant female animal. It causes fetal hemorrhage at a low dose in rats. This is increased synergistically in the presence of its breakdown product, 2,4-dichlorophenol. Synergism occurs when the combined effect of two chemicals is greater than the sum of their separate effects. That is a study done in the Soviet Union. It's the only study of fetal synergism that I've found. In other words, the question hasn't been asked in U.S. research. 2,4-D caused malformations and fetal death in the animals only at a high dose, but it caused various kinds of malfunction and growth retardation at very, very low doses.

"Cancer takes an average of twenty years to develop in humans. The latent period is shorter in children. The younger the animal the shorter the latent period. It can be quite short in children...

"One of the commonest misconceptions perpetuated by the chemical industry is that there's a safe level of a carcinogen—that a little bit won't hurt you." says Dr. Shearer. "With a cancer initiator, that's definitely not true. The changes it causes in cells are irreversible. These changes are not repaired, and they are cumulative over the lifetime of the person. The idea of a 'safe' amount of a carcinogen is a fallacy."

* * *

In the early 1960s, the United States began its first defoliation missions in Vietnam, using a mixture of 2,4-D and 2,4,5-T called Agent Orange[2]. By 1964, reports of illness and damage in South Vietnam came from observers and doctors there (although they were dismissed officially as "Viet Cong propaganda"). Following the spraying of 2,500 acres in 1964, a Saigon doctor reported:

At first [the people] felt sick and had some diarrhea, then they began to feel it hard to breathe and they had low blood pressure; some serious cases had trouble with their optic nerves and went blind. Pregnant women gave birth to stillborn or premature children. Most of the affected cattle died from serious diarrhea, and river fish floated on the surface of the water belly-up, soon after the chemicals were spread[3].

During the year the Saigon doctor wrote his report, Bionetics Laboratory (a subsidiary of Litton Industries) was conducting experiments for the National Cancer Institute to determine the mutagenic, carcinogenic, and teratogenic potential of widely used pesticides and industrial chemicals. The studies were to be evaluated by a blue-ribbon commission of independent experts.

By the summer of 1965, Bionetics studies indicated that both 2,4,5-T and 2,4-D caused birth defects in mouse offspring at small doses that were not lethal to the pregnant mothers. A preliminary report of these studies was completed in 1966, but Food and Drug Administration officials and other agencies did not learn of it until 1969[4].

How a government-sponsored study of such import could be suppressed for so many years remains a puzzle. The White House apparently feared that disclosure of the Bionetics report would encourage international criticism of American chemical warfare in Vietnam and feed growing antiwar sentiment at home. Dow Chemical Company, the major manufacturer of 2,4,5-T and 2,4-D, also applied pressure to suppress the report5. For political and financial reasons, vital health and safety information about chemicals applied heavily both at home and abroad was withheld from the public. In Vietnam, in Oregon, in Arizona, and in other places where people suffered predictable effects from phenoxy herbicides, those who complained were dismissed as propagandists or uninformed hysterics by officials who knew—or should have known—how valid those complaints were.

Industry at least was well aware of some effects of their products on human health long before the Bionetics study was commissioned. In March, 1949, at a 2,4,5-T factory owned by Monsanto Chemical Company in Nitro, West Virginia, an accident released a quantity of unknown "intermediate chemicals" involved in the steps of 2,4,5-T production. Two hundred twenty-eight workers developed a severe skin disease known as chloracne, an eruption of the skin of the neck, back, and chest into a mass of ulcerations, comedones, cysts, and pustules[5].

Chloracne has been repeatedly described by the chemical industry as "only a slight skin rash."[6] Their euphemistic description does not indicate other more serious symptoms of the disease, including urinary disturbances, liver damage, leg cramps, shortness of breath, intolerance to cold, loss of sensation in extremities, demyelination of peripheral nerves, fatigue, nervousness, irritability, insomnia, loss of libido, and vertigo. The condition can persist or recur for many years[7].

Chloracne was not a new disease. It had been known since 1899 in factories producing chlorine by hydrolysis. It was a common affliction during World War II among workers in plants producing chlorinated naphthalenes. In 1936, over three hundred workers in Mississippi contracted severe cases of chloracne after treating lumber with a Dowicide formulation of tetrachlorophenol, used as a wood preservative. In Dow's own plant in Midland, Michigan, in 1937, twenty-one workers suffered from the disease.

With the development of 2,4,5-T and related phenoxy compounds, large-scale commercial production brought outbreaks of chloracne at factories worldwide. Clearly the disease was caused by at least several different chlorinated compounds. In 1957 the most potent of these compounds was identified as 2,3,7,8-tetrachlorodibenzo-p-dioxin (TCDD), an impurity formed under heat and pressure in the manufacturing process of 2,4,5-T and other phenoxy compounds.

The use of herbicides as chemical warfare agents in Vietnam increased the demand for phenoxies, straining production facilities and leading to several outbreaks of chloracne. In 1964, over seventy workers contracted the disorder at Dow's Midland, Michigan, plant[8].

Dow officials became concerned and sought to determine how much TCDD humans could be exposed to without ill effects. In an apparent effort to measure dose-response relationships, Dow contracted with a University of Pennsylvania dermatologist, Dr. Albert M. Kligman, to perform secret experiments with TCDD on prisoners at Holmesburg Prison in Philadelphia. (Kligman was conducting similar experiments with other chemicals for the U. S. Army at the time, testing to determine the maximum doses of chemical-warfare agents required to mentally disable 50 percent of a population.)

Between 1965 and 1967, Kligman applied varying doses of TCDD to the skin of seventy prisoners who were not informed of the identity or possible effects of the compound. At least eight of the men developed chloracne during the experiments, and effects on the rest are unknown. No follow-up studies were done to determine the fate of the men after their exposure to the chemical. No record was kept of the "volunteer" prisoners' names[9].

(It is important to note that studies on the effects of TCDD on humans involved only men, and that industrial exposure apparently only involved male workers. The likelihood that this extremely potent chemical might have severe effects of a different nature on females was evidently not considered; yet industry scientists are quick to dismiss out of hand symptoms reported by women from sprayed areas.)

By the time of the Hart Senate hearings in 1970, both government and industry were well aware of the awesome properties of the TCDD (dioxin) component of 2,4,5-T. They were also well aware of the toxic effects of 2,4-D, the most widely used phenoxy herbicide, which was thought not to contain

TCDD. Over the years of 2,4-D use, many ill effects on humans and animals had been reported and confirmed by scientific studies: swollen eyes, mouth, and lips, rashes, urinary disturbances, renal damage, fatigue, nausea, vomiting, anorexia (inability to eat), diarrhea, swelling and pain in extremities, paralysis, ventricular fibrillation (serious irregularities in heartbeat), neuritis, paresthesia, numbness in legs, fingers, and toes[10].

"Fairly soon after they are oversprayed or spill 2,4-D on themselves, or are otherwise exposed," Dr. Shearer reports, "both animals and people develop weakness, with stiff muscles and rapid fatigue, and show signs of injury to the central nervous system. Of course there is also the usual nausea, vomiting, headache, irritability—symptoms which are hard to pin on a particular cause but always follow toxic exposure to 2,4-D."[11]

That 2,4-D could also cause birth defects was established in the Bionetics studies. The Mrak Commission – a blue ribbon committee appointed by the Secretary of Health, Education and Welfare to review the effects of pesticides – analyzed the Bionetics report and listed both 2,4-D and 2,4,5-T along with five other pesticides under the heading, "These should be immediately restricted."[12]

In spite of such findings, only 2,4,5-T attracted the concern of the government, whose cursory restriction was quickly appealed by Dow Chemical Company in 1970. At the time, 2,4,5-T was not widely used on food crops, but 2,4-D was and still is. Not surprisingly, 2,4-D production has been many times greater than 2,4,5-T[13]. The economically least significant chemical on the commission's should-be-immediately-restricted list—2,4,5-T—conveniently contained a highly toxic contaminant, TCDD, upon which years of controversy would focus while its sister compound, 2,4-D, continued to be produced and used freely.

As the herbicide market expanded, new systemic compounds were developed and joined 2,4-D, enjoying immunity from the ill repute attached to TCDD and 2,4,5-T. Research and regulatory action were centered on TCDD, a red herring that effectively drew critical attention away from the known effects even of purified phenoxy herbicides, whose component elements alone should have suggested their toxicity.

The synthetic hormones known as phenoxy herbicides are all derivatives of chlorophenols, which are varied combinations of chlorine and phenol. Chlorine and phenols had been in use by the medical profession for a century before they were combined by researchers attempting to synthesize plant growth hormones. Both chemicals had revolutionized medical techniques in preventing infection. Chlorinated lime, introduced by Ignaz Philipp Semmelweis as a scrub for doctors and midwives, ended the scourge of childbed fever that killed 25 percent of hospitalized mothers in nineteenth-century Europe[14].

Phenol, or carbolic acid, was first used as an antiseptic by Joseph Lister, the father of modern surgery. Armed with this new weapon against infection, Lister developed techniques of urinary-tract and cancer surgery and amputation previously impossible in medicine[15]. That the biocidal properties of these two "friends of medicine" would be enhanced in combination is not surprising. There is strong evidence that both benzene (the chemical base for phenol) and chlorine cause cancer in humans.

Herbert Henry Dow (1866-1930) pioneered techniques for producing synthetic phenol from bromo-benzene which were later adapted to a method using chloro-benzene. His techniques established his company as the largest producer of synthetic phenol in the world. (Phenol is a basic ingredient of several explosives, as well as pharmaceutical and pesticide products.) Dow based his company originally in Midland, Michigan, because of the locality's deposits of bromine-laden brine, from which he extracted bromine and chlorine.

The chlorination method of producing phenol led to the development of the chlorophenols, from which are derived the phenoxy herbicides (2,4-dichloro-phenoxy acetic acid; 2,4,5-Trichloro-phenoxy acetic acid). Dow Chemical Company became a major producer of these herbicides[16].

Phenoxy compounds mimic the behavior of their analogue, indoleacetic acid (IAA, or auxin), the natural hormone that controls plant growth and metabolism. In plants, IAA controls such functions as the development of branches, the direction of growth, and the ripening of fruit. With extraordinary potency, the phenoxy herbicides accelerate IAA mechanisms to a fatal degree, disrupting nearly every biological activity of the plant: water, mineral, vitamin, and oil balance, photosynthesis and carbohydrate production, respiration, nitrogen and phosphorus metabolism, and enzyme structure and activity. The plant dies from any one or combination of these effects[17].

After 40 years of extensive study, no one knows exactly how phenoxy compounds exert such influences over life processes. What is known is that phenoxy herbicides profoundly affect nucleic acid metabolism, disrupting synthesis of RNA and DNA and suppressing synthesis of gene-regulating enzymes. (In other words, they scramble the genetic code that directs both cell division and cell function.) The result is a wanton proliferation of cells, particularly in the stem, forming swollen, tumorous masses of cells which disrupt the metabolism and "circulatory" system of the plant[18].

Phenoxy herbicides also disrupt production of adenosine triphosphate (ATP), the primary compound responsible for storage and release of energy within the cell. In addition, they destroy the membrane integrity of root cells, resulting in softened tissues and decay[19].

What is important about all these effects is that they involve mechanisms vital not only to plants, but to animals as well. Gene structure, energy regulation, and cell membrane permeability are essential to all life. These mechanisms developed in the plant kingdom long before animal life appeared on earth to inherit them from their plant precursors. Although some of these biochemical mechanisms came to have functions in animals that were quite different from their function in plant metabolism[20], others survive essentially unchanged in cells of all organisms, from bacteria to man. Chemical poisons or radioactive elements which can affect these basic processes do not respect the elusive dividing line between plant and animal kingdoms.

Chemicals that affect genetic mechanisms cannot be expected to discriminate between plant and animal cells. In the union of a single microscopic egg and sperm (or pollen grain and ovum in plants), the structure of the new organism and the future of its offspring depend on the precise alignment of two frail sets of chromosomes coming together in a recombinant "dance" first choreographed in blue-green algae over a billion years ago. Along each chromosome, coiled double strands of nucleic acid (DNA) "unzip," separate, and replicate. Arrangements of four chemicals (nucleotides) in the strands spell out a code transmitting genetic information and instructions for development of the organism. The fertilized egg divides, repeating a mystery as old as Genesis.

This process and the helical molecules of DNA are similar in all living organisms, whether virus, bacteria, primrose, redwood, goldfish, hawk, or man. All living cells share their origin in this primordial marvel of reproduction. In the single cell born of egg and sperm, the same process spells what our children will become. Poisons that affect this process violate the biological and spiritual basis of creation, and are cause for serious concern.

Molecular biology had not developed techniques for studying the effects of chemicals on genetic mechanisms in the 1940s. Safety standards for pesticides at that time did not consider such effects. The first safety tests for 2,4-D reported no immediate toxic effects in sheep and cows grazed on treated pastures. In one test a cow was fed 2,4-D for three months, and Dr. Kraus himself, who pioneered the research on its potential as a crop-killer at Camp Dietrick, claimed to have consumed one-half gram 2,4-D per day for three weeks with "absolutely no effect." On the basis of such "experiments," the U.S. Department of Agriculture cleared the chemical for marketing in 1945, triggering profligate growth in manufacture and use of 2,4-D and related compounds. Research was intensified to find other systemic herbicides for use in agriculture[21].

In the early 1960s, as was mentioned earlier, the publication of Silent Spring and the monstrous birth defects associated with the sedative drug thalidomide alerted the public and the government to the need for more thorough testing of drugs and chemicals[22]. Thalidomide demonstrated that chemicals producing little or no effect in a parent could profoundly influence the development of an unborn child. Scientists realized that substances which affected the structure of an embryo might operate by altering the genetic behavior of cells. Researchers studying the mechanisms of cancer recognized a connection between substances that caused birth defects and substances that caused cancer and mutations[23].

Key to the understanding of this connection is an understanding of mutations, which are permanent changes in chromosomes or their component genes. Such changes—breakages, translocations, additions, or deletions of genetic material—can occur in any cell. In germ cells such as egg and sperm (or cells that produce egg and sperm), such changes affect the structure, growth, and metabolism of developing offspring. In somatic (body) cells, mutations and other abnormal genetic changes can alter cellular development and metabolism. These changes can be transmitted to other cells in the organism, producing cancerous proliferations.

Since even a single molecule of a gene-altering substance can affect genetic processes, there can be no safe level of exposure to these substances[24].

Birth defects, cancer, and mutations are readily identifiable tragedies which a reasonable society would make every effort to prevent. But chemicals which fatally affect the basic life processes of plants may have other less obvious effects on animals and humans that, in the long run, may be equally tragic.

The phenoxy herbicides cause behavioral changes in people and other animals exposed to them. In humans, these changes can be profound, involving impaired memory and concentration, disturbances of sex drive and potency, depressions, fits of temper, hypersensitivity to light and noise, sullenness and irritability, fear and anxiety, alienation. Such effects have persisted for many years after exposure[25].

In addition, phenoxy compounds affect the future behavior of animals exposed to them before birth. Apparently, the chemicals cause changes in brain chemistry which are transmitted from one generation to the next. These changes are expressed in altered patterns of learning behavior in experimental animals[26]. Just how the herbicides cause these inheritable behavioral disturbances is unknown. Chemicals which disrupt particular life processes in plants may interfere with animal biochemistry in very different, unpredictable ways[27].

Indoleacetic acid (IAA), for example, which directs growth in plants, is involved in a very different biochemical mechanism in animals. Tests show changed levels of the neurotransmitter hormone serotonin, both in animals exposed to phenoxy herbicides and in their offspring[28]. IAA is involved in the biochemical production cycle of serotonin, and it is possible that the phenoxy analogue to IAA interferes with that cycle[29]. (Serotonin levels are also affected by stress, and changes in serotonin levels may also be the result of stress to the animal caused by exposure to a poison.)

However the phenoxy herbicides cause these effects, the implications are serious. Serotonin is important to sleep-wakefulness patterns and regulation of behavior[30]. Abnormal levels of serotonin (both high and low) have been linked to spontaneous abortions[31], migraine headaches[32], depression[33], and mental retardation, particularly in conjunction with certain birth defects, such as Down's syndrome (mongolism)[34].

The new science of behavioral teratology—the study of behavioral birth defects—reflects growing concern about the behavioral and psychological effects of toxic chemicals[35]. The association of herbicides with altered serotonin levels, particularly in the offspring of exposed animals, is at the very least an indicator, a warning flag, marking unknown and largely unstudied problems.

Critical Mass

On Christmas day of the year our children were sprayed with herbicides near our farm in the Five Rivers Valley, my husband and I read a news story that appeared simultaneously in various Oregon newspapers. All the articles apparently were drawn from the same press release. The articles dealt with herbicides in forestry, quoting extensively from Professor Michael Newton, an Oregon State University scientist who claimed expertise on the subject.

We were familiar with Newton's scientific work from previous news stories about his use of surplus Air Force Agent Orange stocks on private land in our area. His project had been halted, and EPA had confiscated the remaining barrels of Agent Orange[1].

The article that appeared in our local county paper on December 25, 1975[2] was typical of others which appeared in Portland and Eugene papers. Newton, describing himself as a "forest weed ecologist," claimed to have been studying the effects of herbicides since 1958. He reviewed the "necessity" to control forest "weeds" such as alder and maple in order for conifers to grow, and dismissed any suggestion that herbicides posed a threat to public health or wildlife. Citing his participation in a National Academy of Sciences study team that investigated reports of human health effects and environmental devastation caused by U.S. use of Agent Orange in Vietnam, Newton declared, "Most of the upland forests in Vietnam were destroyed by fire, not by 2,4,5-T...Typical symptoms of 2,4,5-T and dioxin in people were also lacking."

The articles renewed our distress of the summer before. Those statements by an "expert" presented an incomplete, misleading picture of herbicides to a trusting public. We had always been somewhat shy and had not thought of writing to newspapers before, but those articles were the last straw. We pulled out our boxes of studies and wrote a response. Only one paper printed the letter, our local county weekly:

The Newport News-Times, January 8, 1976
Dear Editor:

Re the article on Michael Newton and herbicides, which appeared in the December 25 (1975) issue of your paper.

Professor Newton may find it easy – and no doubt profitable – to like herbicides, living and working safely in the Willamette Valley, where little is sprayed. Perhaps he would like to move his family to the Siuslaw Forest for the summer, where 15,000 acres are scheduled for defoliation this year, 1976-77. Professor Newton claims to have been "studying the effects of herbicides since 1958." If that is so, he would do well to improve his study habits. We have been studying the effects of herbicides since last spring and summer, when the helicopters sprayed the forests surrounding our valley, and have made the following observations:

1. Ten percent of the chicks, 6 percent of the goslings and 17 percent of the ducklings hatched on our place were hatched with various abnormalities, which included crossed beaks, malformed legs and feet, small body size at hatching, and retarded growth.

2. Our children—in otherwise excellent health—suffered severe headaches accompanied by bloody noses and nausea during and after the spraying.

3. Our vegetable crops—tomatoes, beans, and grapes especially—suffered various degrees of damage, and we lost most of the tomato plants altogether.

4. The spraying was done with little regard for wind and climate conditions, one application being applied just before one and one-half inches of rain

Our interest in the subject aroused, we did some research at OSU, where Professor Newton is employed, and discovered the following:

1. That present formulations of 2,4,5 T and the levels of dioxin they contain, used at present rates of application, are highly toxic and that there is no safe dosage for them.

2. That 2,4,5-T and its contaminant dioxins are teratogenic to humans, i.e., capable of causing severe birth defects at almost immeasurably small doses, in the parts per trillion category. Possible defects include: cleft palate, abnormal leg positions, kidney abnormalities, decreases in average fetal weight, intestinal hemorrhage, umbilical hernia, spina bifida; to these must be added the dangers of maternal weight loss and cancer of the liver. Such effects have been noted with "pure" 2,4,5-T, as well as with varying proportions of dioxin.

3. That dioxins accumulate in the food chain, reaching concentrations as high as 9,222 times the original dose in mosquito larvae, one of the primary foods of ducks and fish, especially salmon, in coastal waters. This could present hazards to humans, particularly pregnant women, who consume meat from wild game and domestic animals which grow in these areas.

4. That according to other members of the National Academy of Sciences, who accompanied Professor Newton to Vietnam, as well as members of the Herbicide Assessment Commission of the American Association for the Advancement of Science (1970, 1971), stillbirth and malformation rates increased (particularly spina bifida and cleft palate) with intensity of spraying in Vietnam. Other studies, coming from North Vietnam, carefully document a number of cases of children

with various malformations born to women in sprayed areas and a number of primary carcinomas of the liver traceable to the same source. This is in direct contradiction to Professor Newton's assertion that "typical symptoms of 2,4,5-T and dioxin in people were also lacking."

We who live and work in these forests and suffer directly from the effects of the spraying receive no financial support to fight these practices. Indeed, our tax money supports the very persons, such as Professor Newton and the Forest Service, who are using us as guinea pigs in an experiment designed for their own profit.

The readers of your newspaper, especially those who live, work, hunt, fish, or visit in the national forests of this state, deserve to know what is being done to them in the name of "forest ecology" and "vegetation management."

Sincerely,
Carol and Steve Van Strum

* * *

Publication of this letter in The Newport News-Times had unexpected results. People phoned and wrote from all over Lincoln County, as well as from neighboring counties of Lane and Benton: a spray-truck operator who still suffered kidney damage and skin problems six years after being forced by ill health to quit his job; a hunter accused of killing five elk who was acquitted after autopsies showed they had died from herbicide poisoning; farmers who had lost cattle and had to destroy horses which had gone lame and lost all their hair; tree planters who had become sick after planting sprayed units and had found numerous small animals dead soon after spraying.

Several families reported being poisoned by eating berries from sprayed roadsides. A logger called, upset about the number of blind fawns he had seen over the past few years. "Their eyes are all shrunk up," he said, "like the eyeballs and sockets just didn't grow, and they bump into things. If the doe moves a little too far from them, it's all up. They don't make it very long, anyways." He wondered if the herbicides could be causing the blindness. We didn't know[3].

Many of the callers refused to identify themselves, for fear of losing their jobs in the timber industry. Women were less restrained by this fear. Two housewives in Five Rivers, Susan Parker and Susi Gilbert—both mothers of small children—called a meeting in the local two-room schoolhouse to discuss the letter.

The meeting was attended by a large cross-section of the valley population. Old people came who had lived in the area all their lives, who had seen the helicopters spray the hills and watersheds year after year without ever knowing or questioning what their government was doing. There were farmers, loggers, millworkers, and several Hoedads, members of a tree-planting

Image by Todd Moore.
Fisher Schoolhouse, the site of the first meetings in the fight against herbicide spraying

cooperative who had first-hand experience with herbicides on the units they were hired to replant. People braved twenty miles of near-impassable logging roads to come from Deadwood, a community to the south of Five Rivers. The room was full.

Spread on the table were a collection of photocopied studies and the Forest Service E. I. S. Here were gathered the neighbors we had lived among for nearly two years, and we recognized almost no one. Steve explained what had prompted our letter to the paper, and presented a brief summary of our research: the origins of the phenoxy herbicides during World War II, the effects reported from their use in Vietnam, the laboratory tests resulting in birth deformities in test animals, the studies that showed bioaccumulation of TCDD in the food chain.

Of course there was no proof, he said, that what happened in a laboratory could happen here in the woods, but there was also no proof that it couldn't happen.

Perhaps the children's illness, the deformed chickens, ducks, and geese, and the dying garden crops were all a coincidence. Perhaps the dead fish and ducklings in the river, the crayfish washed up on the shore, the dead songbirds were also a coincidence. He had gone to the library seeking assurance that it was all a coincidence, and instead found strong indications that it was not.

The people in the Forest Service and road department and the people at EPA who regulated the chemicals, do not live in the forest, he said. But we do. We work here, fish and hunt here, grow our food, hike with our kids. From day to day we see the deer, the elk, the hawks and fish, the trees. We know their habits and the cycles of their lives, and we are in a position to recognize better than anyone when something has gone wrong. We wrote the letter because we had seen something go very wrong.

"You have all lived here longer than we have. What have you seen?" we asked.

Every person in the room had a tale to tell. Beekeepers had lost their bees after the spraying. A rancher had lost 23 out of 36 young heifers after the road through their pasture was sprayed. (We learned later that it was our pasture those heifers died on; it had been leased to the farmer the year before

Image by Rio Davidson
Old growth natural forest in Valley of the Giants, North Fork of the Siletz River

we moved there.) One woman had had fourteen miscarriages in the years she had lived in the valley. Another told of her two miscarriages, and of her son born with defective lungs and liver. The young wife of a logger had been unable to complete a pregnancy in the five years they had been married. An elderly couple told how their health suffered so badly every year during the spraying that they moved out of the valley for those weeks; they had always thought that they alone suffered such a problem.

Fishermen and hunters told of deformed fish and deer, all agreeing that the aquatic life of the river had declined drastically. The crayfish, fresh-water mussels, salmon, trout, and steel head populations were but a fraction of what they were ten or twenty years before.

Almost every family could report a case of intestinal, respiratory, or nervous-system problems, cancer, miscarriage, hemorrhaging (frequently resulting in hysterectomy) in women, chronic illness among the children—all in an area remote from industrial pollution or urban traffic.

Hearing their own experiences confirmed by the reports of neighbors was gratifying, but the implications were frightening for everyone. If the herbicides were causing all these problems, what could be done? Who among us would have the temerity to challenge the federal government?

"If it were my neighbor, now, or even a timber company making my kids or cattle sick." one man said, "hell, it wouldn't make no difference how safe they said it was, those helicopters would never get off the ground. They're easy enough targets, after all. But this is the government we're talking about..."

Beyond the fear of reprisals and intimidation by the government was the more practical and tangible fear of losing jobs. The families in the area all depended, directly or indirectly, on the timber industry for their livelihood. The timber companies who hired them were in turn dependent on the Forest Service for their contracts, and no one had the slightest doubt about the fate of anyone who stirred up trouble against the Forest Service.

Steve suggested that fighting the Forest Service might not be necessary. Perhaps no deliberate evil was involved. The Forest Service is handed the latest forest management tool from agricultural researchers, he said. The Forest Service depends on the EPA to assure the safety of the chemicals, and the EPA depends on the safety studies done by the chemical companies whose ultimate responsibility is to their stockholders who know nothing about the chemicals. It was entirely possible that the Forest Service didn't know about the studies we had found, or what happened in the valley every time they sprayed.

Susan Parker, president of the Five Rivers Organizing Group (FROG), suggested everyone write down their experiences and opinions about the spraying, anonymously if they wished. Their notes could be collected and incorporated into a letter to the Forest Service. She passed out note paper and pencils. Assured of anonymity, everyone in the room wrote detailed reports. While they wrote and talked among themselves, John Noell, who had come to the meeting from Deadwood, introduced himself to Steve. He explained that he had a scientific background and offered to help with any further research. John's "scientific background" included a Ph.D. in biology, but at the time it wasn't something he took great pride in.

John is a tall, exuberant man in his mid-thirties with a boisterous enthusiasm for everything from music and mechanics to spiders, newts, and molecular structures. Kids love him. He's always ready to explore something new, or look up answers to hard questions. On a rescue mission recently to help a neighbor whose car was stuck on the "divide"—the high ridge between the Five Rivers and Deadwood Creek watersheds—John reminisced about the 1976 meeting.

"I was new to the area. I came to the meeting out of idle curiosity, really. I listened to Steve's presentation and my interest was aroused, against all my firm intentions to avoid such things. He was good. He was calm. He wasn't out to destroy anybody. He had his facts under control, and he was using them not so much to prove a point as to raise questions. I liked that. It's an element that was lacking in the science I knew.

"I was a real drop-out. Until that meeting, I hadn't the slightest desire to look back. The scientific community seemed to me to be a microcosm of all that was wrong with the world—pettiness, ego-trips, competition, self-indulgence, a perpetual compromise of honesty or integrity. It was all a game, you had to play by the rules. I was good at it, but I came to see that what I was good at was destroying me, undermining my life, my family.

"So I dropped out. We moved to Deadwood, bought a small farm, and I set to work learning to be a farmer. You know the vision—a small farm, the idyllic life, good clean work, a warm fire at the end of the day, watching your children thrive on fresh healthy food. And then the bubble bursts.

"I don't mean the vision is unreal, it's just irresponsible. There is no dropping out, not so long as you live on the planet. Helicopters come, shatter your peaceful world, and your first instinct is to run away, find a little corner of the earth where these things don't happen.

"Well, there are no more corners to hide in, and what would be the good? Even to be able to run away is a luxury others can't afford. And with DDT in Antarctic penguins, PCB's in the tissue of every American, radioactive wastes dumped wherever they can get away with it—well, there's no place to run to, anyway. And even if there were, who could enjoy sitting there watching the rest of the world die?

"I wasn't thinking of all that when I met Steve, though. One thing for sure—I had no idea what I was getting into! All I realized at the time was that the threat he described was real. I was more than familiar with the chemicals in those studies.

"We had one child then. We were planning to have more. That's what I thought of first—I don't want that stuff anywhere near my kids, either the one we have or the ones that aren't born yet. If a dog attacks your child, you don't really stop to think, you just grab the nearest weapon available to beat it off. That's the way I felt then, and I grabbed the nearest weapon—my training in science—and joined the fight.

"I had been disillusioned with science, because of the nature of the game. It's called the game of who's better, and the winner is the one who can demolish the credibility of all comers. It's a game I'd been good at, which is nothing to be proud of. But if the battle was to be fought on that ground, I had the right weapon for it.

"And unfortunately that's the ground the battle is fought on. Not just with chemicals, but nuclear power, dam construction, oil tankers, pipelines, bombs, everything—it's all left to the experts to decide. You end up with courts of experts arguing the fine points of issues that if carried to their logical conclusion are sheer lunacy—in biological terms, anyway.

"What it comes down to is how much life one is entitled to destroy for a profit. It's called risk/benefit analysis. To play the game you abandon at the start the basic question whether any life is worth sacrificing merely for a profit. To bring that one up disqualifies you as emotional and unscientific. People are taught to leave such matters to the experts, trust the experts, delegate all responsibility to the experts, let the experts decide what is right and wrong, good or bad.

"What impressed me about that meeting was seeing Steve—this businessman-turned-farmer—explaining the principles of molecular biology and chemistry to an audience of loggers and housewives, and they were hanging on every word. They were getting a look at what all those 'experts' had been up to!

"I've thought a lot about that first meeting, and the events that followed, up to the federal court hearings. We had no idea what we were getting into. We had no precedents to follow. We had no intention more grandiose than protecting our own families, and no certainty whatever about our methods, yet we accomplished something. It didn't seem like much at the time, but it mushroomed. It's still growing.

"How did it happen? A bunch of strangers with completely different histories, knowledge, personalities, and interests were brought together by a common threat. It triggered a kind of critical mass phenomenon..."

* * *

"Like a lot of people, we moved here to get away from things," a beekeeper said at that first meeting. "My God, surrounded by National Forest—what could go wrong?"

The notes written by people in the schoolhouse that evening testified that much had gone wrong. Susan Parker read their comments aloud, and we drafted a letter to the Forest Service that everyone signed. The letter outlined all the problems people reported and requested the local ranger district to send representatives to a meeting called for March 4.

Frank Rasmussen, the district ranger for the Alsea Ranger District of the Siuslaw National Forest at the time, attended the March 4th meeting, along with Jim Warner, who directed the district spray operations, and Gene Klingler, the district silviculturalist. Except to acknowledge that the Forest Service had been somewhat lax in notifying people about what they were doing, the three men did not seem to take seriously any of the community's concerns.

Klingler did most of the talking. He emphasized that the Forest Service relied on EPA safety rules, that the herbicides had been approved by the EPA, and that it was not up to them as Forest Service employees to decide on the safety of their sprays. When asked directly if he could cut back on the spray plans around the Five Rivers valley, Rasmussen said it was not up to him to decide. The ranger and his men left the meeting without committing themselves to any change in their program.

The people of the community were left in bewilderment.

"We just sat there for a few minutes staring at each other." one farmer-logger recalls. "I guess you could say we were pretty innocent. It was a big thing to do, getting together and writing that letter and meeting with the government. I don't think this community ever got together on anything before like that!"

He scratches his head and lifts his eyebrows, a fleeting mime of Stan Laurel. "Everyone was so surprised to find out everyone else felt the same way about this thing. I guess we felt the Forest Service couldn't help feel the same way, too, once we told them what we knew. It was a shock, don't you see, to do this thing that was such a big step for us and then they didn't even care and wouldn't listen. It was a blow, and it kind of destroyed our faith, if you know what I mean. Nobody knew quite what to do next."

Having taken such a step, the community was unwilling to give up. A committee, including residents of both Five Rivers and Deadwood, was nominated to investigate what could be done next. The following week, the committee met in Deadwood and chose a name for the organization that would represent the two communities in their efforts to stop the Forest Service from spraying their watersheds. The name was Citizens Against Toxic Sprays (CATS). The same committee formed the officers of the group, who made several trips to Eugene during the next few weeks, trying to find a lawyer who could advise them.

"None of us knew anything about law or the administrative procedures open to us under Forest Service rules." Susan Parker recalls. "If we had known then what we know now, maybe things would have gone differently. At the time, though, it was a question of playing by their rules and keeping our resistance legal, or letting people handle it their own way. Out here, that means rifles. Knowing you're right gives you confidence—maybe false confidence—that you can win, even in court. But we certainly had no idea what we were getting into!"

In addition to serving as FROG president and as an officer of CATS, Susan was still nursing her infant daughter and working with other Five Rivers families on a model of the local covered bridge to enter as a float in the annual "Beachcomber Days" parade in Waldport, that June. Our letter to the newspaper had interested her immediately because she had had surgery to remove cancerous growths from her breast since moving to the valley, and both her daughter and her husband, a mill worker, had serious respiratory problems whenever spraying occurred.

"Finding a lawyer isn't as easy as it sounds." she says. "We discovered that there aren't many who know anything about environmental law, and even fewer who know anything about chemicals. They do know a lot about money, and the first few we went to either were interested only in whether or not they could make a bundle out of the case or just couldn't understand the issue."

Divorced now, Susan teaches and practices massage therapy on the coast, the single parent of a son and daughter. She is also chairperson of the Lincoln County Planning Commission and is still active in citizens' groups opposing herbicide use. "Once you get involved, it's hard to stop." she says. "We were so naive in '76. We were so discouraged, we'd about given up on lawyers for the day, when we walked into Bruce Anderson's office. We didn't know anything about him, but someone had told us he'd sued to get a power line moved out of the way of ducks migrating or something, and we thought we'd take a chance. He listened to us for about an hour, then said to come back next Monday. He would take the case.

"We went home feeling high, but worried, too. Getting a lawyer means getting serious, and also means getting money. I have to say this for Bruce. He stuck this thing out without the remotest chance of getting his money's worth, even though he was charging us only half his usual fees. And he had this respect for what we, the CATS folks, knew and could learn. It was a two-way effort—he took advantage of our science, and we took advantage of his law. Both sides learned a lot."

Bruce Anderson remembers fondly that day in 1976. "A group of people from Deadwood and Five Rivers came in to see me and told me some things that were very upsetting to me personally. The problems they were having were serious, and I thought maybe I could help."

On the wall opposite Bruce's desk in his Eugene office are photographs of Bruce, his wife, and their two children and dog, hiking, backpacking, skiing, mountain climbing, fishing. He smiles, recalling over 600 hours of legal work devoted to the CATS case by himself and his assistant, Doug du Priest.

"I had no idea that day that it would ever turn out to be war that it did." Bruce says. "None of us could ever have realized the magnitude of it. The more I learned about the issue and talked to people who had been poisoned, the more I realized the immensity of the chemical problem in this country. Even with personal concern and commitment, it was frustrating; without them it might have been impossible. I experienced the limitations on somebody in a small, private practice being able to involve themselves in multiyear battles, defending people and the environment against the combined resources of industry and government."

Since Deadwood and part of Five Rivers are located in Lane County, Eugene newspapers had taken an interest in the herbicide issue from the beginning. Urban people surprised rural members of CATS with their interest and support. Many had been distressed for years over county use of herbicides on roadsides, parks, and school grounds, and consulted CATS for information and advice. They also contributed money, badly needed for legal fees and other expenses. Rural residents of other forest areas also joined CATS, and circulated petitions among their neighbors.

On April 19, 1976, twelve days after Bruce Anderson agreed to take the case, he submitted a proposal to the supervisor of the Siuslaw National Forest. The proposal was modest enough. It asked only that herbicide spraying be suspended in portions of the Siuslaw National Forest within a mile of the Deadwood and Five Rivers communities, and offered suggestions for experiments with alternative manual methods of controlling brush. Four days later, the Forest Service turned down the proposal.

The only recourse remaining was federal court. Money was the greatest obstacle, but an awakening public responded to the news with donations, bake sales, and benefit concerts, supplying enough funds for Anderson to proceed.

It was spring, and the helicopters would be flying any day.

On May 12, CATS, joined by Oregon Environmental Council and the Hoedads treeplanting cooperative, filed suit in federal district court in Eugene, asking for an injunction against spraying of 2,4-D, 2,4,5-T, and 2,4,5-TP (silvex) throughout the Siuslaw National Forest, on grounds that the Forest Service EIS on herbicide use was inadequate.

"That surprised a lot of people, suing a whole forest just on the grounds their fat environmental statement was 'inadequate,'" Susan Parker said. "Inadequate is such a wishy-washy term. What really surprised us was to learn there is no law you can invoke to stop your government—or anyone else—from spraying poison all over you. All you can do is say they didn't prepare an adequate statement telling what the poison would do! And the whole thing had grown so big—it wasn't just a bunch of families out in the woods pleading with one district of the Forest Service any more. It had got out of our hands, involving the whole USDA, federal court, chemical and timber companies. That was frightening. I guess they meant it to be."

Cathy Noell acted as accountant for the growing CATS group. John Noell, Steve Van Strum, and Stephen Hager—an oceanographer-statistician from Deadwood—prepared the scientific basis for the court case and arranged for expert testimony from scientists throughout the country.

Cathy soon found herself working almost full-time to keep CATS' books current, as donations came in and expenses rose.

"We weren't prepared for this at all." she said, displaying a large kitchen table covered with account books and records piled between trays of apricots for the fruit dryer. "We were asking for so little to begin with—just for them to keep their spray away from our farms. They forced us to go higher and higher up the ladder. Every step of the way we would have settled for quite a bit less, and every step of the way the other side escalated."

Anticipating a lengthy court proceeding, CATS members were still faced with the spraying program scheduled for that spring. Early in May, they appealed to federal Judge Otto Skopil for a temporary restraining order to suspend spray operations until the case had been decided. Judge Skopil denied the request, as Forest Service contracts had already been signed and to break them would be unduly costly.

Immediately, the Forest Service proceeded with its spray operations in the Siuslaw. One irate farmer from Five Rivers followed the Forest Service crew over miles of rough logging roads to the helicopter landing, his loaded deer rifle poised visibly on the dashboard.

"Those arrogant sons-a-bitches, that was my water supply they were going to spray and they knew it," he said. "They knew I signed that letter and an affidavit for the judge too. Maybe I wouldn't of shot them, but I sure as hell intended to put the fear of God into 'em. It shook 'em up some. They radioed ahead and the helicopter took off before we got there. They didn't spray that unit—not that day, anyway."

Publicity on the case inspired widespread resistance to county road spraying in the Siuslaw, particularly in Lane County. Lincoln County Commissioner Jack Postle, who had halted the use of herbicides by Lincoln County, attended a public meeting of the Lane County commissioners in Eugene to urge them to follow his example. So long as there was any doubt about the safety of the herbicides and their effects on human health, he could not justify exposing people to them, he said. Against his advice, Lane County continued with its roadside spray program.

Shortly after that meeting, John Noell stopped a Lane County spray truck as it passed up the road past his house and farm, spraying directly into the water supply for his house and garden. John met the truck armed with his most formidable weapon—a sample bottle—which he ordered the sprayers to fill with the mixture from their tanks. He took a number of samples of water from his water supply below the sprayed part, in numbered bottles. One of these bottles he spiked with a heavy, measurable dose of the spray mixture.

"I complained to the county road department and submitted my samples to them for analysis." John said. "I gave the samples to the head of the road department, who sent them to the chemistry lab at OSU for analysis. Dr. James Witt was in charge of that, and as I recall it was a technician named Marvin Montgomery who was actually running the tests. So those sample bottles passed through several people's hands, you have to understand.

"Now I spiked that one sample really heavily. There's nothing underhanded about that, it's a standard practice in analytical sampling. You spike one or several samples with a measured amount, and that's how you know your equipment is accurate. But here's the kicker—the report they sent back stated that all the samples were negative.

"You can look at it a number of ways. Either the head of the road department dumped out what was in the bottles and refilled them before they went to the OSU lab, or someone there at the lab refilled them. Or if you really want to be nasty, you could suspect them of never running the samples at all and just manufacturing the test results."

John might have made more of the incident, except that Cathy was pregnant at the time. Several weeks after the Noells' water supply was sprayed, she was hospitalized with a particularly difficult miscarriage.

"That hurt," John says. "I got into this thing partly because it concerned our life on this farm, partly because it appealed to the scientist in me, the old challenge of the game of experts. But we really wanted that baby, and losing it changed the game. It wasn't a game anymore. It was life or death. It was too painful a thing to exploit."

Time was short for preparing the amount of testimony needed. John was working with Steve Van Strum and lawyer Bruce Anderson, preparing the case for hearings in federal court on June 23 and 24. Judge Skopil permitted witnesses to submit written testimony in lieu of personal appearances in court, saving time and expense for expert witnesses located across the continent. Only witnesses for cross-examination would be required to appear.

The Forest Service had been joined by the Industrial Forestry Association (IFA) as interveners. The IFA was represented by a law firm previously engaged by Dow Chemical Company, when the company announced its intention to intervene in the case[4].

The anxiety of preparing the case was compounded by small, harassing incidents. That a lawsuit over herbicides could be important enough for such tactics seemed absurd at the time, but the incidents mounted.

Steve and John were often "tailed" on the 80-mile drive to Anderson's office. The same two men with a camera seemed to appear wherever Steve went, both in Eugene and on the coast. A reporter for a small coastal newsletter was contacted by Dow representatives offering information and a trip to Midland. Forest Service employees frequently visited unannounced at our home, whether we were there or not (our doors had no locks, and we had never worried about that before). Telephones began behaving oddly, with erratic noises, static, and outages unrelated to weather. FBI agents appeared in Deadwood and Five Rivers, asking questions about CATS officers.

Dow cameramen equipped with audiovisual equipment attended the Lane County commissioners' hearings on roadside spraying, filming, and recording all who were present. The same crew, along with a Dow public-relations officer, visited our home unannounced shortly before the federal court hearings. They were met at the driveway by the Forest Service district manager. The incident prompted Bruce Anderson to call the federal attorney, threatening to charge the government defendants with litigating outside the court.

"You don't want to believe it—you don't want to live under that kind of paranoia." John Noell said. "But I think if you're not paranoid, you're just not looking!"

"Too Many Unanswered Questions"

On June 23 and 24, 1976, the CATS case was heard by Judge Skopil in a special session of district court held in Eugene, Oregon.

Weeks of intense legal and scientific preparation were about to meet the test. CATS officers suddenly were faced with an unexpected crisis. No one possessed clothing appropriate for the grandeur of federal court. John retrieved from his archives the corduroy jacket he had worn in his "Dr. John" days. He came to court smelling like mothballs.

Steve made an emergency call to his father in California, who located a pre-1930 pin-stripe suit of his own and sent it by U.P.S. The suit hung in the living room, where the children guarded it fiercely from pet birds, dogs, and two orphan turkeys that had been adopted by our parakeets. They called it the "Law Suit."

The public benches of the district court were filled on the morning of June 23. Housewives, farmers, and loggers who could take a day off came from various parts of the Siuslaw Forest. CATS supporters from the Eugene area added to the crowd. Work clothes and "country best" dress contrasted vividly with the uniform somber business suits of government witnesses and observers. Steve Van Strum and John Noell sat with attorney Bruce Anderson and his assistant, Doug du Priest, at a table in front of what one rural child present called the "altar rail" dividing the public benches from the court. To their right at another table sat the federal attorney and attorneys from Industrial Forestry Association. Between the two tables and the judge's bench sat the court reporter, a pale, extraordinarily dead-pan gentleman who sat motionless during the proceedings, only his fingers moving, flickering silently over the keys of his stenographic machine. Most of us had never been in federal court before and were surprised to see that the judge wore a black robe.

CATS had submitted massive testimony to the judge on the toxicity and environmental fate of phenoxy herbicides, both 2,4-D and 2,4,5-T, from scientists all over the United States. Siuslaw Forest residents submitted affidavits telling of damage to health, crops, land, and livestock from forest spraying.

Other witnesses testified to the damage inflicted on the crop trees – conifers—whose growth the chemicals were intended to promote. Gerry Mackey, president of Hoedads, had submitted detailed testimony on the feasibility of manual brush control as an alternative to herbicides, and other forest workers testified concerning the effects of forest spraying on their health. No CATS witnesses had been called by the defendants for cross-examination.

CATS had called a number of Forest Service witnesses for cross-examination, particularly the scientists from Dow Chemical Company, the OSU, the EPA, and the Council for Agricultural Science and Technology who had supplied most of the information in the EIS in question. On at least one occasion an OSU scientist did not recognize either the title or text of a scientific paper published in his name.

In a handsome, handmade wood file box, oceanographer Stephen Hager had organized a complete file, cross-indexed, of all available references and relevant documents supporting or refuting the government's scientific studies. At opportune moments, Hager—a bearded, stocky figure in overalls—surfaced from the back row of public benches with his box, to hand Bruce Anderson a relevant document, annotated with questions for the witness. CATS was able to impeach many industry claims on the spot.

On several occasions, defense lawyers objected to Judge Skopil's continuing references to them as "the Dow lawyers." The judge seemed to find this amusing.

Only one CATS witness appeared to testify in person at the hearings, called not by the defendants for cross-examination, but by CATS under subpoena. This was Dr. Patrick O'Keefe, a chemist from Dr. Matthew Meselson's lab at Harvard. Meselson and his associate Robert Baughman had pioneered an analytical technique for detecting and measuring TCDD at levels of parts per trillion (ppt)[1]. These are astronomically small amounts, but for a chemical as deadly as dioxin, such measurements are vitally important. Dr. O'Keefe had successfully refined Meselson's and Baughman's technique to reach even lower detection limits, as low as one ppt[2].

Meselson and Baughman had found a range of from 18 to 814 ppt of TCDD in fish and crustaceans caught for human consumption in Vietnam waters after massive defoliation missions there[3]. (Using techniques developed since 1976, Dow studies have shown multigenerational reproductive effects from TCDD at levels of 10 ppt per day on a body weight basis; EPA interpretation of the same Dow studies found such effects at one ppt per day[4].)

The finding of TCDD in the food chain in Vietnam was significant to CATS' case, as the Forest Service EIS had flatly stated that no evidence had ever been found to suggest that TCDD bioaccumulates in food chains[5]. Dr. O'Keefe's work with Meselson in refining analytical techniques for TCDD was performed under contract with the EPA's Dioxin Monitoring Program. One of the questions CATS asked Dr. O'Keefe was whether or not any TCDD had ever been found in biological samples from the United States.

Dr. O'Keefe replied that samples from the U. S. had indeed been analyzed, and that some of the samples analyzed under the Dioxin Monitoring Program had been taken from the Siuslaw National Forest and had contained high levels of TCDD. The samples had been collected for EPA by Dr. Logan Norris of the Forest Service Pacific Northwest Forest and Range Experiment Station at OSU[6]. Dr. Norris's work had been referenced in the EIS, and he had testified in behalf of the Forest Service on his work studying the effects of herbicides on forest animals. He had collected the Siuslaw samples in 1973, had known of the positive analysis reports, yet did not submit the information to the Forest Service for consideration in the EIS. Only negative analysis results "were considered sound," Norris said.

The animal samples Norris had collected from forest and roadsides in sprayed areas of the Siuslaw were mostly small mammals and songbirds. Those which showed the highest concentrations of TCDD were wrens, a Stellers jay, and deer mice. TCDD was also found in small animals and birds from other Pacific Northwest forests, and some very high levels (up to 485 ppt) of TCDD had been detected in shrews from roadside areas in Virginia[7]. These were all creatures high on the food chain, omnivores or carnivores whose tissues might accumulate high concentrations of persistent chemicals present in their food.

"What puzzles me is that no game animals were tested," another CATS witness commented. Had the researchers been looking for evidence that TCDD bioaccumulates in the food chain, they should have been looking at the game animals that humans consume. Human exposure is supposedly the primary concern of such testing. Why the researchers did not collect grouse or quail, or take tissue from deer and elk, along with the salamanders, shrews, wrens, and other creatures taken for analysis, was never explained.

Dr. O'Keefe's testimony was electrifying. As the only CATS witness to appear at the hearings, he presented an impressive contrast to defense witnesses. Their answers to Bruce's questions were categorical and evasive, stoutly proclaiming the safety of the herbicides even when presented with compelling evidence to the contrary and confronted with contradictions in their own work.

Within the narrow semantics of law, Dr. O'Keefe alone of the persons who took the stand that day qualified as a witness in the truest sense. TCDD was more to him than an issue to be weighed by countless research papers. TCDD was something he had found, measured, and observed personally. He answered questions simply and directly, with a genuine concern for making clear to a layperson the nearly mystical complexities of his equipment and techniques for measuring unimaginably small quantities of a chemical. He told what he knew and with equal candor told what he didn't know. He drew no conclusions, took no sides. None were needed.

It was startling testimony. It shattered the Forest Service contention that TCDD had never been shown to bioaccumulate. The Forest Service had said on several occasions that if herbicides or dioxins ever were found to be accumulating in animals, they would not use them. Yet, Dr. O'Keefe's testimony showed that it had been found, right in our own forest, two years before.

Dr. O'Keefe testified as well about the production of dioxins from burning 2,4,5-T, a long-standing concern of forest residents because most of the sprayed vegetation is subsequently burned. He also discussed EPA studies that found dioxins in beef fat from cattle grazed on sprayed rangeland in Texas.

A memo from Dr. Ralph T. Ross, then coordinator of the Dioxin Monitoring Program, had reported the preliminary results of the beef-fat studies in August 1975[8]. Residues of TCDD in various samples of the beef fat ranged as high as 60 ppt. We had attached this memo to our response letter to the Forest Service EIS.[9]

The Forest Service had printed the letter in the final draft of the EIS, but had not printed the enclosed beef-fat memo on the grounds that it had not been intended for publication[10]. They claimed that the EPA findings were relevant only to "pioneering analytical methodology," although the memo concluded, "Studies including teratogenic and other toxicity effects indicate that the residue levels [in beef fat] mentioned above may present a health hazard to man based on the application of normal margins of safety."

Dr. Ross, the author of the memo, appeared at the hearings as a defense witness. His testimony dismissed the beef-fat memo as inconclusive. He reiterated the Forest Service arguments and dwelt at length on the scientific debate over methods of detecting and measuring TCDD.

"For many of us this came as a surprise," Susan Parker said. "It was hard to believe that this was the same person who had written that memo, with its warning about health hazard to humans. He seemed to have done a complete about-face. But then, since he wrote that memo he had changed jobs and was working for USDA."

CATS and other people generally felt that EPA itself had acted more as adversary than as ally in the herbicide controversy. In the courtroom a number of witnesses for the Forest Service had served as advisers to EPA. Testimony in the case revealed that while EPA had been reassuring alarmed citizens, the agency had been sitting on information about dioxin in Siuslaw animals and U.S. beef. The information never was released voluntarily by EPA but was only made public through subpoena and fortuitous circumstances. EPA was well aware of the significance of its data, yet allowed public exposure to TCDD to continue while scientists quibbled—not over whether TCDD was there, but over how to measure it most accurately. In spite of the beef-fat data, 2,4,5-T use is still permitted on rangelands.

The federal court hearings continued for two days in June without all witnesses appearing. Judge Skopil recessed the court until August 10, 1976.

During the intervening weeks, Dr. O'Keefe, whose testimony had been vehemently attacked by the defense, particularly with regard to the Siuslaw Forest samples, reran those samples using his improved analytical technique and was able to confirm the findings of TCDD[11].

During his first visit to Oregon in June, Dr. O'Keefe had arranged to collect human milk samples from women in the Siuslaw Forest, which he tested along with samples taken from nursing mothers in herbicide-sprayed rangeland areas of Texas as part of the EPA Dioxin Monitoring Program. One sample from a nursing mother in the Siuslaw National Forest proved to contain approximately one ppt of TCDD. This news also was not released by EPA until Oregon Congressman Jim Weaver requested the test results under the Freedom of Information Act and presented them to the public[12].

The Forest Service, OSU, Dow, and Industrial Forestry Association witnesses challenged the milk sample results even more vehemently than they had the forest animal data. "Me thinks the gentlemen protest too much," one nursing mother commented.

"They always said they would stop spraying if they ever found dioxin accumulating in the food chain," said Steve Tedrow, a mill worker who raises cattle in Alsea. "But that's meaningless if they refuse to believe it when someone actually finds it. You never hear them doubting any negative results, ever—even negative results by the same scientists who find the positive ones."

The defense also maintained that analytical testing of fish and shellfish from Vietnam waters and human milk from Vietnam was not relevant to CATS' case and needn't be mentioned in the EIS, although Dr. O'Keefe had rerun Vietnam samples and confirmed the presence of high levels of TCDD[13].

Two witnesses for the government testified that the Vietnam data were not significant in terms of forest use of herbicides because the rates of application and the levels of TCDD present in the Agent Orange used in Vietnam were much higher than in the U. S. But Dr. George Streisinger, a molecular biologist at the University of Oregon and a CATS witness, presented strong evidence that differences in dosages used in Vietnam and in the Siuslaw Forest were far less than the defense claimed[14].

The two defense witnesses who testified on Agent Orange use in Vietnam had both been members of the National Academy of Sciences committee which prepared a study of the effects of Agent Orange in Vietnam. One defense witness, Dr. Michael Newton, whose press statements in late 1975 had prompted our letter to the local newspaper, was still remembered for his unauthorized application of surplus Air Force stocks of Agent Orange on private timberland within the Siuslaw Forest[15].

Dr. Fred H. Tschirley, the other defense witness on Agent Orange, had (along with Newton and others) co-authored a report on phenoxy herbicides for the Council for Agricultural Science and Technology, an industry-sponsored group that published "studies" exclusively in support of agricultural chemicals[16].

Judge Skopil was not convinced that Agent Orange was irrelevant.

"The failure of the 1976-77 EIS to mention the Vietnam controversy is unreasonable, particularly when spraying of 2,4,5-T in the Siuslaw National Forest is frequently carried out near the homes and farms of local residents," he wrote in his decision. He added in a footnote, "Several local residents testified that they had suffered physical symptoms similar to those reported in Vietnam."[17]

From the testimony presented, the judge concluded:

> The 1976-77 EIS fails to acknowledge the extreme toxicity of TCDD or the opinions of scientists about its hazards, to discuss EPA administrative proceedings against 2,4,5-T or the ongoing TCDD residue monitoring program being conducted by EPA, to report on TCDD levels found in animal specimens taken from the Siuslaw National Forest, or to mention the controversy over the effects of Agent Orange in Vietnam. These are the major, but by no means the only, shortcomings of the EIS in its discussion of the potential effects of phenoxy herbicides upon human and animal health.

Judge Skopil's decision to grant a permanent injunction against the use of 2,4,5-T and silvex (2,4,5-TP) was also based on the inadequacy of the EIS in discussing alternatives to the use of herbicides in forestry. Gerry Mackey, president of Hoedads, had pointed out numerous inadequacies in the figures quoted in the EIS. The costs of herbicide use presented in the EIS were far below actual costs reported by a Forest Service silviculturalist, and the costs of manual brush control were inflated far beyond actual costs.

In addition, the judge noted, the EIS contained:

> ...no discussion of the relative effectiveness of herbicides and other methods of vegetation control in enhancing long-term productivity. It provides no data on the acreage now being treated by each method. It does not discuss possible combinations of the different methods employed for vegetation management or possible variations in methods based on geographic or other differences between National Forests. It does not indicate whether other herbicides may be substituted for the ones recommended or provide information on how effective such substitutes might be. In short, the EIS does not rigorously explore or objectively evaluate the proposed herbicide program and the alternatives to it. It cannot, therefore, serve adequately either those who must employ it to make decisions about vegetation management in the National Forests or those outside the decision-making process who wish to evaluate the alternatives.

In the judge's opinion, however, the inadequacy of the discussion of alternatives was inseparable from the inadequate discussion of health issues presented by TCDD-contaminated herbicides, as the consideration of alternatives to herbicides was important chiefly in light of the risks presented by TCDD. His decision therefore related only to 2,4,5-T and silvex, and not to forest use of 2,4-D.

Judge Skopil made it very clear that his decision was based on the Forest Service's blatant lack of regard for the consequences of their spray program on human health. "No subject to be covered by an EIS can be more important than the potential effects of a federal program upon the health of human beings," he wrote.

His ruling ordered that "plaintiffs are entitled to a permanent injunction enjoining further applications of the TCDD-contaminated herbicides 2,4,5-T and silvex in the Siuslaw National Forest unless and until the defendants have properly remedied the defects in this EIS." In other words, the Forest Service could not spray 2,4,5-T or silvex in the Siuslaw Forest until they had prepared a new EIS that included all the information on health effects and alternatives found to be lacking in the old EIS.

During the years the CATS case was in progress, reports from other countries concerning toxic chemicals—many of them involving herbicides and dioxins—further dramatized the potential dangers of poisons capable of affecting people's lives without their knowledge and consent.

- The day before the second CATS hearing in August, 1976, news broke of a chemical disaster in Seveso, Italy, in which a factory explosion released a cloud of dioxin onto the village and surrounding countryside[18].

- In New Zealand, three mothers of babies born with spina bifida called for an investigation of 2,4,5-T and a moratorium on its use until more was known about its effects. Physicians writing in the New Zealand Medical Journal called on New Zealand doctors to examine possible connections between herbicide exposure and neural-tube birth defects (including spina bifida and anencephaly, or absence of a brain).[19]

- Studies from six hospitals in Sweden found a dramatic increase in severe congenital malformations in infants born to hospital workers who used hexachlorophene soaps[20]. (Hexachlorophene, like 2,4,5-T, is derived from trichlorophenol, and contains dioxin[21]. Hexachlorophene soaps such as Phisohex were used routinely to wash newborn infants in this country until recently, when scientists discovered the soaps to be linked to severe brain damage in children washed with them at birth.)

- Scientists in Vietnam reported a five-fold increase in liver cancer in areas of Vietnam sprayed with Agent Orange. Reports of birth defects, including mongolism and central nervous system defects, had been made even during U. S. defoliation operations in Vietnam[22].

- By 1977, U.S. veterans of the Vietnam war were beginning to report symptoms similar to those suffered by the Vietnamese and by exposed populations in the U.S.[23]

In Alsea, Oregon, Bonnie Hill, a school teacher and mother, read the information packet CATS had assembled and began to wonder if miscarriages suffered by her and other local women were related in any way to heavy spraying of herbicides by the Bureau of Land Management near her home in 1975[24].

* * *

Judge Skopil's opinion was filed on March 7, 1977, effectively canceling the spray program for that year. The Regional Forester for all national forests in Oregon and Washington, anticipating further lawsuits based on this precedent on other national forests, suspended spray programs for all national forests of the region[25]. Nationally, Forest Service use of 2,4,5-T plummeted.

There was great rejoicing among CATS members and citizens of the rest of the Pacific Northwest forests, but most knew that the relief was only temporary. "The Forest Service will just come out with a new E. I. S. for next year's spray season, stuff enough science in it for the judge to pass it, and go on like they always have," said Cathy Hankins. "We gained a year—one spring and summer without helicopters. One year of relief, and it cost us $24,000. If we were millionaires it would be worth it, but none of us are, and we still owe the lawyer half that money."[26]

The Hankins' water supply and land surrounding their pastures had been sprayed repeatedly by the Forest Service over the years. Cathy's husband Mike had submitted an affidavit in the CATS case describing illnesses and deaths among livestock and wildlife following the sprayings[27].

"You can't stop them for long, and they won't stop themselves," Cathy said. "They'll include all that stuff about dioxin in their new EIS—how poisonous it is and how it's getting in the beef and mothers' milk—and go right ahead and use the stuff. So long as their EIS is legal next time, the decision is up to them. No one can stop them. And they're so damn arrogant they'll never admit there's anything wrong with the stuff."

In April 1978, the following year, Cathy's prediction was confirmed. Judge Skopil, having reviewed the new EIS for the year's spray program, reluctantly acknowledged that it fulfilled his requirements in adequately presenting information on health effects of herbicides and discussion of alternatives that had been lacking in the old EIS. After noting that he was compelled to limit his review to the issues and deficiencies ruled upon in his prior decision, Judge Skopil qualified his decision:

> My decision that the new EIS deals adequately with the defects found in the prior EIS does not mean that the disclosure was as complete as it could have been. The Forest Service has chosen to give greater emphasis to studies that find little or no danger from these herbicides. Adverse information is not as prominently discussed as information which favors the use of phenoxy herbicides. The choice of language for describing certain studies and reports of illness and injury seems to downplay the importance of this information.
>
> Although I have decided in favor of the Forest Service in this case, the Forest Service should understand that materially misleading statements and conclusions can be just as fatal to an EIS as omissions and false statements. Also, perhaps the public outcry against the use of these herbicides would actually be reduced if the Forest Service were more candid about adverse information in the EIS[28].

Judge Skopil reviewed the limited role of the court in judging not the consequences of an agency's actions but only whether or not the agency had taken a "hard look" at those consequences. As a judge, Otto Skopil could say no more, since the Forest Service had complied with the letter of the law. But as a human, moved by considerations beyond law and procedures, he did say more, entering into the record his own quietly impassioned plea for the Forest Service to act responsibly in its decision-making:

> The E.I.S. discloses that there are unanswered questions about the potential harm from a chemical that is acknowledged to be one of the most toxic substances known to mankind. If I were the person responsible for making the decision of whether herbicides containing this substance should be broadcast sprayed from helicopters over our national forests,

I would be extremely reluctant to allow it. There are simply too many unanswered questions, and too few benefits compared to other methods of vegetation management, to justify the unknown risk...I am sincerely concerned about the effect of the use of herbicides upon the health of human beings and our environment. I sincerely hope that the Forest Service will review all new information and will re-evaluate its decision.

The Forest Service, however, read only that it was now legal for them to proceed with herbicide spraying, and triumphantly loaded their helicopters for another assault on the Siuslaw.

A Soldier Turns to Home

Return? The soldier turns to home
To find it moved, and walks –
Like poets – unguarded through the trees.

—Alan Swallow

From a helicopter at three or four thousand feet, the rain forest below seems to be in motion, rather than the aircraft. It is not a forest of trees, branches, trunks, soil, and water, but a variegated cloud of green, billowing past. Rivers and creeks are shining ribbons blurred by steam rising in early morning heat. Ahead and above the rotor, giant thunderheads tower thousands of feet high.

Seen from the ground, the helicopter is an awkward, unlikely insect clacking its way to the horizon. To its passengers, though, the only reality is noise, wind, incessant shaking, and mountainous puffs of green streaming past below. The loud, flapping "Whock, whock, whock" of the blades beats relentlessly on eardrums and brain. Every turn of the rotor shakes the metal shell of the machine with bone-jarring rhythm.

A hot wind pours through the open cargo doorways, drying the film of grit and oil deposited at take-off on skin, hair, and clothing.

Below, the green forest canopy is broken occasionally by large river valleys, flat bottomland, doll-house hamlets, the crazy-quilt patterns of rice paddies. Trucks, jeeps, and motorcycles move like toys threading the landscape of Vietnam's Binh Dinh province.

Late in 1969, a young man crouched beside a 1,000-watt trumpet loudspeaker mounted on a frame that had been lifted into the helicopter. Beside him, another man readied boxes of leaflets, and just behind them, crew chief and gunner manned the machine guns on either side of the craft. The pilot brought the helicopter low over a small village at the mouth of a valley where two rivers met. A single two-story cement building dominated a cluster of mud and thatch houses. Buildings, paths, palm trees, banana trees, and dirt roads were contained within a concertina-wire perimeter, separating them from surrounding rice paddies.

As the helicopter circled the village, the young soldier swung the hinged speakers out the cargo door and tinkered with the control panel of a tape recorder and amplifier. Grasping the strut of the seat, he leaned out the open doorway to throw the leaflets his partner handed him, while the loudspeakers blared their message in Vietnamese, in a voice clearly audible 7.6 kilometers away.

"The Viet Cong have been saying the sprays make the people sick and are making them sad, and that the sprays are killing the elders and the children. The Viet Cong are lying. They are poisoning the water, so the people will believe their lies. The Viet Cong do not like the sprays because they let the soldiers see where they are hiding."

The young man was Paul Merrell, the Psychological Operations Combat Loudspeaker Team leader. Twelve years later, from the small farm in Five Rivers where he now lives with his five-year-old son and two adopted children, Paul recalls that loudspeaker message and the twenty-seven months and one day he spent in Vietnam. Like most Vietnam veterans, Paul came home from war to face a more insidious and demoralizing kind of combat – the constant battle for health and self-esteem.

"The first three years I spent in and out of VA hospitals. In between I would hide in my father's house, sitting in the basement and hitting the floor at the slightest noise. There were long spells where I couldn't eat, couldn't keep anything down except Coke and popcorn. I would try to work, tried going to school, but it always fell apart. I would get sick, get violent, so shaky I couldn't stand up. I tried getting married, but that fell apart, too, and I found myself with a baby to raise. A wreck raising a baby – it was ludicrous."

In 1978, Paul plunged almost unwittingly into another war when he became involved with growing national resistance to domestic use of the same chemicals that had broken his health in Vietnam. "It was another kind of fight altogether, fighting battles in court and in public meetings, doing research, and learning and teaching others at the same time. The only effective weapons were knowledge and communication.

"It saved my life. It brought everything into focus for me, and this time there was no doubt what I was fighting for – for other veterans much sicker than I was, for our right not to be re-exposed to the same poisons in wheat fields or roadsides or forests; for my son, for my friends, and for the places that were the heart and soul of my childhood – the trout streams and mountains where my dad and I used to fish.

"That was what snapped something in me to begin with. They were going to spray White Sand Creek..."

Almost overnight, Paul became a kind of dervish in the anti-herbicide movement in the Northwest, feverishly educating himself and anybody who could keep up with him in the fields of chemistry, politics, law, and bureaucracy. With Georgia Hoglund of Citizens Against Toxic Herbicides in Clarkston, Washington, he established a communications network between veterans' groups, environmental organizations, and citizens groups nationwide. A printer by trade, he printed and distributed his Toolkit, a resource manual providing access to scientific documents, government agencies and regulations, case law, legislation, Freedom of Information Act procedures, organizing techniques, public speaking – everything he had learned in the battle to keep his beloved forests of northern Idaho from being sprayed[1].

By 1978, when Judge Skopil made his final ruling on the Siuslaw Forest EIS, concern over forest spraying had spread throughout the Pacific Northwest. Citizens' organizations to fight forest poisons sprang up in northern California, southern and central Oregon, Washington, and Idaho. CATS, exhausted and still heavily in debt from its court battle, could not continue to meet increasing demands for scientific, medical, and legal information. In early 1978, the Northwest Coalition for Alternatives to Pesticides (NCAP), based in Eugene, relieved CATS of this burden. By 1982, NCAP had grown into a coalition of over a hundred community-based organizations, organized under four regional councils. (In 1981, CATS—the group spontaneously formed to fight a single court battle—quietly dissolved, its debts still unpaid.)

Citizens Against Toxic Herbicides (CATH), based in eastern Washington and northern Idaho, became an active and effective element in NCAP. As a link between CATH and other NCAP member groups, veterans groups, and environmental organizations, Paul's travels took him all over the United States. Months on the road, hours on the phone and at the typewriter, alternated with bouts of sickness and exhaustion for several frantic years.

In 1980, Paul and his son Zack moved to Eugene to work with NCAP as network coordinator. Five Rivers, 80 miles from Eugene in the coastal mountains, became at first a sanctuary, and finally a home for both of them. Here among the forested hills, the farm animals, a constant flow of visitors and children, he carries on his battle by telephone and typewriter, from an almost unnavigable bedroom filled with file boxes, books, stacks of papers, legal briefs, law books, and letters.

"There's something about Five Rivers – the countryside and the people – that I just haven't seen since I was a child. Since we left Kamiah [Idaho] when I was sixteen I never felt I had a home again till I came here. There's a closeness here among the people and the valley they live in. Maybe it's the vultures overhead, maybe it's those alders billowing in the breeze. Maybe it's because there's a bunch of people here who've had enough of the bullshit, who had the balls to take the federal government to court in '76. It's a dedicated, peace-loving community where people don't like the violence of poisons and have gone every legal route to stop them, but would shoot back before they'd see their kids get sprayed again..."

It is spring. The last petals of the cherry blossoms are falling, and apples and azaleas are in full bloom. Two ponies, a horse, and a donkey gallop up a far hill, wheeling and racing down again in some reckless celebration of their own. Paul sits on the hood of his car admiring a makeshift pen full of goslings and Plymouth Rock chicks. He is tan and bearded, wearing a ragged field jacket, with the disheveled, edgy air of combat still about him in spite of his quiet voice and his two rightfooted slippers, picked up at a local garage sale. He cradles a .22 Hornet on his lap, ostensibly watching for crows in the newly

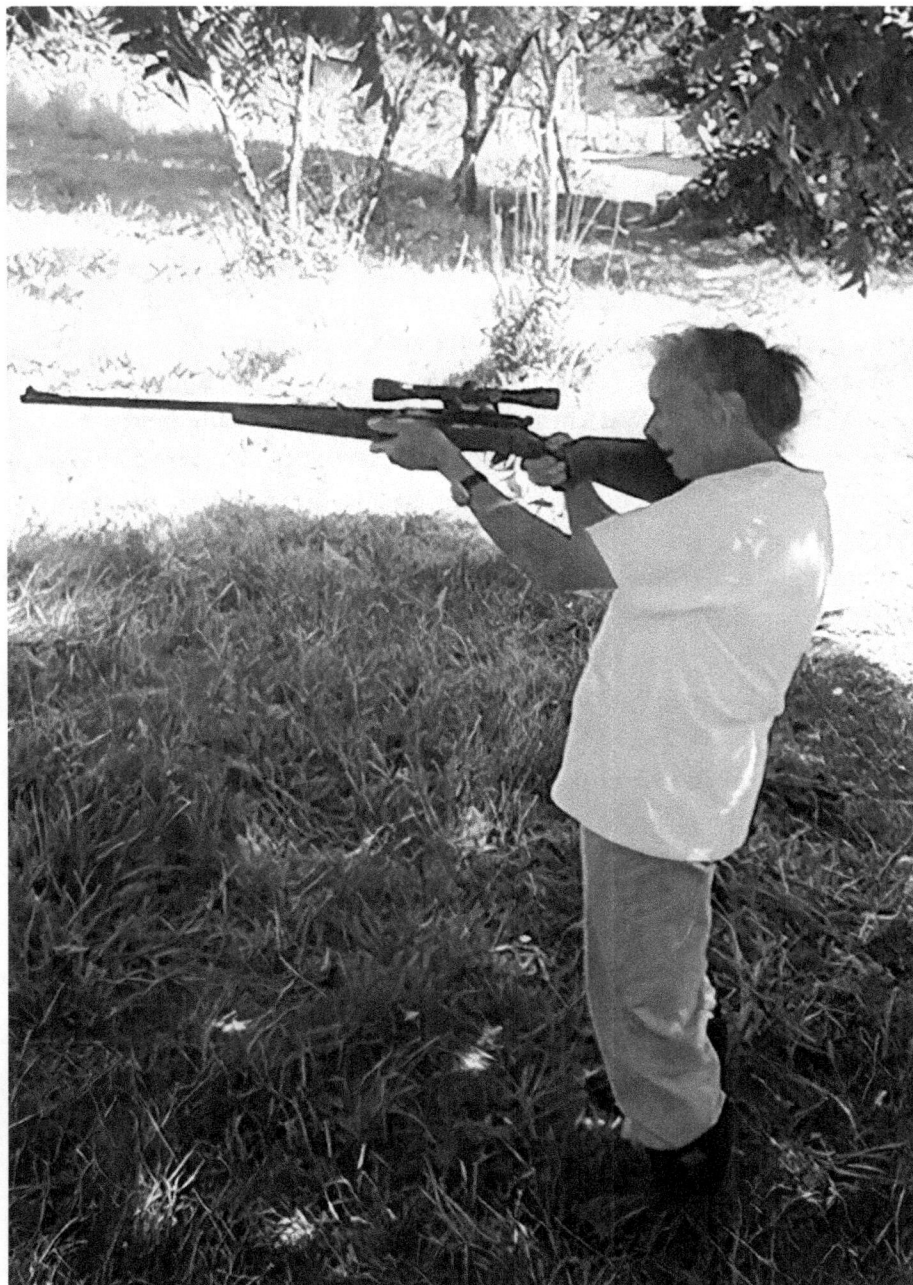

Image by Nikko Merrell
Carol Van Strum keeping watch for varmints on her Oregon farm

planted corn, but he sights the scope on a helicopter passing over a distant ridge, checking to see if it has spray booms.

"It's O.K. Cedar maggots, hauling shake bolts. That's a Huey, a UH-1. We called them Hueys, or slicks. That's what we flew over there. That particular message—the broadcast about the sprays—wasn't our usual line of work. That was in Binh Dinh Province, in coastal Vietnam, about halfway between Saigon and the DMZ. We—the 173rd Airborne—were just holding territory then, not attacking, supposedly keeping the people 'safe' from VC attack or contact. The Army had civic affairs companies in there, too, doing public works, sanitation, propaganda, road-building, that sort of thing.

"What that meant was, they moved the people – civilians – out of particular areas and consolidated them in others. That village we were circling was one of the places they consolidated them, a 'relocation center'. In other words, a concentration camp. Thousands of people, whole families, were removed from their homes – usually forcibly removed – to these centers. The way people lived over there, they lived in small hamlets and villages, often completely cut off from the rest of the world, and farmed the surrounding countryside. They would be moved out and all their crops defoliated with herbicides to keep them from moving back.

"The countryside over there was mostly small valley pockets facing the ocean, with triple canopy jungle on the hills, and in the large river valleys like An Lao would be flat bottomland, rice paddies, small hamlets. In the upper valleys, the VC and NVA were in control. You couldn't bomb them out, there were too many civilians. The pacification program – to win 'hearts and minds' of a population – was a major military and political experiment there. What it meant was, you move the people out, defoliate their land, bomb and artillery the hell out of everything, and spray it. The people were moved downstream to the relocation center. The ones that didn't go got sprayed, and of course the center was downstream from the spraying. There were many complaints about miscarriages, spontaneous abortions, headaches, birth defects, nausea, etc. So they sent us in to broadcast that message.

"Crop denial – to starve out the VC – was an excuse, of course. The real reason to move the people out of the area was to remove the population base for the VC to influence, to get the population centralized and under control

A U.S. Army UH-1 (Huey) helicopter spraying herbicides over Vietnam

instead of out there in the jungle where you couldn't keep track of what they were doing. It struck me as odd at the time that our message didn't justify the crop-denial missions. I never thought much about it, though. That message was completely out of context with the work we normally did, which was with the Chieu Hoi [open arms] Program, to get the other side to defect. The combat loudspeaker team I was in charge of – well, the normal course of affairs for us was to get our asses out where bullets were flying. The spray broadcasts were a vacation from that."

Four or five crows appear over the ridge and circle the garden, cawing loudly. They land in a tree across the river. The rifle lies on the hood of the car, but Paul makes no move to pick it up.

"They know the rules of the game. No one will shoot unless they actually land in the garden. Listen to the buggers – they're jeering." He clasps his hands behind his head, admiring the plowed garden below.

"You'd be flying over the countryside over there and look down on that jungle, the valleys with rice paddies and clusters of trees – banana, bamboo, palm – and the little villages of thatched huts, the jungle round the edges in the mountains. The helicopter would be flying between these huge thunderheads with rays of sunlight coming through them. You could look down and see that mankind can be a positive force on the planet, that the people down there were living in tune with their environment.

"Then you'd see the black spots and gaping wounds in the jungle from bombs, napalm, artillery barrages. The sick yellowish brown of defoliated rice paddy areas, a mess of dead grasses where there had been careful cultivation. You'd come into one of the camps – relocation centers – where the people, who had lived in such a healthy way in that world, were packed together in prefab huts with stucco or cement walls, no glass, tin roofs, all in lines, surrounded by concertina wire and guard towers, the area defoliated, the ground bare dirt or mud, flies everywhere, and everyone carrying national ID cards. Their animals had been killed or abandoned, they were fed on Texas long-grain rice and U.S. food, and there was nothing for them to do but just sit. Women, children, old men, just sitting.

"You couldn't think of these things at the time. You just had to focus on the job you had to do. My job was to get the other side to defect, using broadcasts and leaflets, working from the air or on the ground. We used taped messages or live broadcasts with defectors or interpreters or prisoners. So we were always in the thick of it, we were sent where the VC or NVA were.

"We spent half our time just waiting for something to start, then jump into the helicopter and get there. Then there was the equipment – the speakers and controls were always fucking up. I was always tinkering with them, and when they did work they were so goddamn loud I came home with a profound hearing loss. We'd drop fifty to a hundred thousand leaflets on a mission, one of us leaning out the right side, the other handing him leaflets to throw out and down – away from the tail rotor. That was always a great thrill, leaning out of the cargo door like that. We were always under fire. I went down twice over there, and once we came back from a mission with fifty two bullet holes in the chopper.

"One thing I don't think people realize, even a lot of vets, is that no matter where you were in 'Nam, if you were in the military you were exposed to herbicides, especially Agent Orange. It was used everywhere – along every roadside, around every Allied encampment, along major trails, around landing aprons at airfields, and in vast areas of jungle to improve visibility. The perimeters of major encampments were often dosed with enough defoliants to keep even a single weed from growing. All the major bases were also sprayed – once a week usually – with truck-mounted foggers spraying malathion or carbaryl for mosquitoes.

"There's another thing, too, that a lot of people don't realize, and certainly the Veterans Administration has never considered at all. And that's the matter of evaporation-volatilization-of the herbicides. You don't have to be standing under the spray nozzle to get exposed. The stuff vaporizes for as long as any of it is on the ground or vegetation, and the warmer the air, the more will be vaporizing.[2] And let me tell you, it was hot over there! That's the first thing that hit us when we landed over there – the incredible heat.

"So anywhere we went that had been sprayed, we were breathing that stuff, getting it on our clothes, on our skin, drinking it in the water, eating it in the food. I think you could safely assume that any GIs over there were exposed, one way or another. When I filed for Agent Orange disability with the VA, that's the first thing they asked for – proof of exposure. That's the first thing they ask of every veteran who files for Agent Orange disability – the day, hour, minute, and exact location of your exposure. As if in the middle of getting shot and mortared and firebombed and ambushed you're gonna take out your little notebook and record that you got sprayed or walked through six miles of defoliated forest! And again, you have to remember that no one – not even the guys who were handling the stuff – was warned that it might cause problems, so of course none of the guys bothered to notice when or where they were exposed.

"Our major areas of operations were areas that were technically evacuated. Between artillery bombardments, napalm drops, bombing missions, defoliant runs, airmobile infantry units swept these areas. Wherever signs of enemy or other human activity were found, our 'psywar' team was sent. Sometimes we pulled as many as five or six missions a day, shuttling from one hot spot to another. Usually these areas had been clobbered with everything from tear gas to high explosives to herbicides.

"I think my first major exposure was in Thailand in the spring of 1969. I was smuggled into Bangkok on a diplomatic flight along with nine other members of the Fourth Psychological Operations Group. We were on a classified mission to U-tapao Royal Air Force Base, the big B-52 bomber base, to load leaflet bombs for the first massive bombing strikes in Cambodia. You know, those bombing missions started months before they said they did in the Pentagon Papers.

"U-tapao R.A.F.B. had been specially constructed as the major B-52 base in the Southeast Asian theatre. Its two-mile runway was built on a swamp beside the ocean so the bombers would have no obstacles in their way when they took off. Those fucking jets were so heavily loaded that they had to have auxiliary steam jets to get them going fast enough to take off. They'd drop the

Image by U.S. Air Force
B-52 bomber approaching U-tapao R.A.F.B. after mission over Vietnam in 1972

steam jets in the ocean off the end of the runway. The area that U-tapao was built on was so swampy the base's second largest population was cobras. They were everywhere.

"There was a huge munitions-storage area where hundreds of thousands of tons of bombs were stored, all meant for Cambodia. The bombs were stored in open revetments – a particular pattern of U-shaped mounds designed to keep a single explosion from blowing up the whole base. We were set up in the middle of this area, loading fiberglass leaflet bombs with propaganda messages to clarify the message of high-explosive bombs. The leaflet bombs were rigged with an altimeter device so they would open a couple of thousand feet above the ground, releasing the leaflets. They were dropped along with the explosive bombs.

"The whole munitions area at U-tapao was obviously and totally defoliated, a bald spot in the middle of a tropical savannah. There wasn't a blade of grass anywhere. The cobras would crawl out of the swamp onto the roadways and runways at night to warm themselves on the pavement. In the morning they'd get lost and roam the whole base. In the munitions area we saw them all the time. A lot of them died of dehydration and exposure from lack of cover. U-tapao had one of the widest perimeters of any base I ever saw, and there wasn't a bush growing within a thousand feet of that perimeter. I don't remember seeing any birds, except seagulls. The enlisted barracks were built on ground that had no vegetation, just sand. There was so little growing – here you are in the tropics, in the jungle – and you just fried. There was no cover from the sun.

"The first symptoms I noticed were a gut-ache and my sinuses plugging up. I remember getting so nauseous it affected my balance. I was dizzy all the time. I got horrible diarrhea. My shit was black with blood. My urine turned cocoa-brown and burned horribly. I got this incredible headache – not your oh-god-I-better-take-an-aspirin type headache but a full-fledged axe-between-the-eyes variety. My legs almost became worthless. They'd buckle as I walked and I'd end up flat on the ground. My hands and feet went numb. My legs stayed cold, cold all the time, I couldn't get 'em warm. I was running high fever and chills, shooting pains that began in my toes, shot through my whole body and blew out the top of my head. All the joints in my body not just ached – I don't know how to put it-it was pure misery. I was popping 'zits' all around my eyes, my nose, my temples, my ears, the back of my neck, down my back. Coarse, black curly hair started growing on my shoulders and back – for years, my back was almost furry. It was strange hair, the roots were all painful, the damn things would in-grow all the time. My skin got real pasty and would tear real easy. You'd just brush up against something and get a wound out of it. I had sores that would just ooze all the time, a lot of infections.

"But the thing that was the worst was going on in my head. I hit bottom. I've still got letters that I wrote—one to my brother back home—was almost suicidal at that point. Yet here I was in this cushy job, no pressure, no combat. It just didn't figure.

"One of the other guys came down with the same stuff. I went on sick call. They gave me hydrocortisone ointment, kaopectate, and it seems like antibiotics. It didn't really seem to do anything. I was pretty much sick the rest of the two months I was in Thailand. My whole body was just like that, this shakiness. I couldn't understand it. I was like most of the guys in the military, real young, the cream of the crop physically. I'd never been sick before, not like that. And they never could tell me what was wrong. I knew I was sick, I knew at the time that was why I was feeling so low. If they had told me I had malaria or typhoid or the plague or any of the other diseases going around, I would have been overjoyed. I never got an explanation until 1978, when I finally figured out from reading the research what the likely cause was.

"My extension leave – for extending my tour in Vietnam – had been delayed by the mission to Thailand. It was a cushy job, and a chance to visit Thailand, and I'd pulled strings to get the assignment. It still strikes me as ironic that what I wound up doing was pulling strings to get poisoned...

"I came home on leave the summer of 1969 to something that wasn't home any more. My family and friends didn't seem to want to get close to me because they knew I was going back. It was the peak of the antiwar movement, and I still remember people I thought liked me asking me if I was proud to be a baby-killer. And I remember freezing in 100 degree weather and being totally tired and worn out. I overstayed my leave and then went back to 'Nam and was assigned to a provincial propaganda team working out of Da Lat in the central highlands. It was too cold for me. Sometimes it got down to fifty five or sixty, and my hands and feet turned white from the cold. Hell, I was raised in Idaho and always got along fine in really cold weather before then.

"I was sick a lot in the highlands...I pulled strings to get assigned to Binh Dinh Province and from there to the 173rd Airborne—the 'Herd'. I'd had it with the cold country and I'd had it with working around the South Vietnamese Army and drafted U. S. troops. I figured the safest spot for me was with a gung-ho unit like the Herd. I figured if I went back stateside I'd wind up in the stockade because of my attitude towards rear-echelon brass. With the Herd, I was over 300 miles away from my superiors. I was ready to spend the rest of my time in the Army there.

"I was wrong. The sickness came and went. It almost seemed cyclical, a month between peaks. I'd get very ill, slowly recover, and then start getting ill again. Landing Zone English, or LZ English, our headquarters, was typical. A dustbowl surrounded by lush vegetation. The smell of diesel was everywhere. The landing strip was totally defoliated, and I never connected any of it with my health.

"But I do remember one time. Boy, do I remember! It was early spring, 1970. The Herd's AO [area of operations] had turned into a major battleground. We had been on pacification, just holding turf, but then the Second NVA Division moved in on us. The Second was a crack North Vietnamese regular army unit, also called the Yellow Star Division. Things got so heavy that the 173rd wound up getting reinforced by another brigade from the Fourth Infantry Division and by a battalion from the 101st Airborne Division, which was the reaction force for the Second Military Zone. We'd had a couple of months of going night and day, pulling a lot of night reconnaissance missions trying to draw enemy fire so we could locate 'em.

"This one time we got a prisoner who said he was willing to lead us to his base camp. They flew us out in a helicopter to join a company of infantry from the Fourth Division. We were carrying back-pack speakers and were hoping to surround the base camp and persuade the NVA to defect. It was really being gentlemanly to call it defection when you've got 'em surrounded. It worked well with the VC, but rarely worked with the NVA We walked through jungle trails all day and entered a defoliated area that had also been hit by tear gas. Everybody's eyes were running, noses were blistering, and skin was itching from the tear gas. Near dusk we came to a 'yard [Montagnard] rice paddy in a swale surrounded by defoliated jungle. The rice paddy had been defoliated, and the trees and brush around the paddy didn't have any leaves. The only thing still growing was elephant grass. We decided to bed down in the rice paddy. We established a perimeter [posted guards] and bedded down.

"It poured rain during the night. We were dog-tired, and woke up in three or four inches of water. The prisoner couldn't remember where the base camp was, so we scrubbed the mission and radioed for a helicopter to extract us before the sun came over the hill. The chopper was late. When the sun came up and it started getting warm, we all wound up rolling on the ground in pain. It was like thousands of tiny needles stuck in the skin, all over the skin of our bodies. The rain had washed something into that rice paddy. The whole unit moved to a nearby river and bathed with soap. That got it down to where it was tolerable. Then my team left by helicopter. I never heard what happened to the rest of those men.

"In the next couple of days all the old symptoms returned in spades. My nerves were totally shot. I remember hearing them coming to get me for another mission. I remember lying down or falling down on the floor. I dimly remember people trying to revive me. I remember bouncing on a stretcher on the back of a jeep, and I remember waking up at B Med [the medical company] and being told three days had passed while I was out. I had an I.V. [intravenous needle] stuck in my arm. I remember them telling me my white count was way down. I remember them giving me some antibiotics and some tranquilizers and being sent back to duty.

"I pulled one more mission and knew it was my last. I went to Qui Nhon and landed in the 67th Evacuation Hospital. All the same symptoms, even worse. I had myself transferred back to Eighth Psychological Operations Battalion headquarters in Nha Trang. I lasted a couple of days at work and hit the hospital again, Eighth Field Hospital.

"My doctor was Major Ream. I don't remember his first name now, but he and I developed some kind of very close bond. He used to take me out of the hospital for walks on the beach at Nha Trang and for long visits with a Vietnamese family that I was very close to. I recall vividly that Major Ream was very concerned about cancer patterns developing in the civilian Vietnamese population. His time in the military was just about up and he was trying to arrange to come back to Vietnam as a civilian to study the problem.

"My time in Vietnam was just about up, too. By that time I had been in the battalion longer than anyone else. I came back in August, 1970, and lit in the civilian hospital in my home town a few days later. My family doctor had taken one look at me and arranged to get me into a VA hospital because the oxygenation rate in my blood was drastically low, I was extremely anemic and had jaundice, and he said I was a combat fatigue case. I took the bus over to the VA hospital. They gave me a shot when I got there, and three days later I came out of the fog enough to realize I was in the locked ward of a mental hospital, drugged out of my mind.

"Over the next three or four years, I spent at least a year and a half in the hospital. I was so shaky, all the time. All the scars on my body changed color, light to dark. I started getting these great big patches on my body, these dark spots, and long straggly hairs growing like crazy on my back. Warts – things like warts – I don't know what you'd call them, just bumps that are sore all the time. Something that looks like athlete's foot on one foot and in my groin that won't respond to anything. I had that over there. Everybody did – it's hard to find Viet vets who don't have what they call 'jungle rot'. They'd tell us over there don't worry, when you get back it'll all go away. Only it doesn't go away. None of it goes away. My health is better now, but every now and then it all comes back, and with it comes this depression. That's the part I really can't handle, the depression..."

The crows seem to have retired for the day. Paul takes the rifle into the house and returns with an axe. From a pile of alder rounds, he starts splitting firewood for the kids to stack for winter.

"I was lucky, I got really sick before I left 'Nam, and was still sick during the first year after I got home, so they recognized my disability. That's more

Image by US Army

A U.S. Army helicopter used to spray Agent Orange in Vietnam

than most guys get. Half of 'em don't even know what's fucked them up. But I still can't get health care. Oh, I can go to a V. A. hospital, and it's supposed to be free and all that, but I can't pick my doctor. I can't tell them no, you can't send me to a mental hospital. That's been their attitude – if you think you've got Agent Orange symptoms they call you a mental case. I can't get health insurance because of the problems I have, so I'm in this incredible fucking double bind.

"The last time I was exposed to 2,4-D, I hit the hospital with the same old mess of symptoms. Summer of '78, I think. Sitting there in the CATH office with them spraying wheat fields on two sides, and I didn't get out of there. I should have known better, I could smell it and everything, but I still had that little bit of lingering doubt, so I stayed. Next thing I know I'm dropping Zack at some friends' and driving over to the hospital at ten at night because I don't know if I'm going to get up in the morning and take good care of the kid.

Image by USDA Forest Service
A Bell 205 helicopter used for spraying forests in Oregon in 1988

"They ship me over to the VA hospital and the first thing they want to do is load me up with tranquilizers and send me over to American Lake [mental hospital]. And I just said, no, I'm sorry, but you know, I can't do that to myself again. And that's the thing, I can't do it. I can't go into a VA hospital. Never again. I can't let those people—I mean, it's up to me – can't allow them to treat me like that. But it's the same thing now with all these really goddamn sick people that are going into VA hospitals saying, 'Hey, I want to be taken care of, and I think it might be Agent Orange.' They get sent to the psychiatric units, or the VA finally sends them a disability check after they've died of cancer! And won't let his mother cash it because he's already dead – honest to God, that's what happened to Paul Reutershan!³

"It's really hard anyway for Vietnam veterans to accept Agent Orange and all its ramifications, because that's a heavy pill to swallow. You're asking them to swallow high cancer risk, you're asking them to swallow possible genetic effects in their kids, you're asking them to swallow that maybe they really do have some out-of-control emotional problems that they've been denying. You're asking them to swallow some really heavy physical problems they can't look at. It's a pretty hard pill to swallow. It should be. It's pretty heavy..."

The wind that sweeps the Five Rivers valley every afternoon is stripping the last petals from the cherry blossoms, driving them first one way and then another in the cross winds off the ridges. Paul's pile of split wood reaches the height of the chopping block, enough for the kids to stack for today.

"You have to have hope," Paul says, "you have to believe things can be changed, or you might as well stick a pistol in your mouth and pull the trigger. I really hit bottom with a suicide attempt in '72. After that, I threw away their fucking pills. The disability award came through. I kept trying to work. I'd worked full time as a printer for years before going in the service, but after I came home I just couldn't do it. My health would go, my body couldn't take the grind, and then after '75 I was a single parent with a baby to take care of as well.

"Nobody who hasn't been there could possibly understand what it is to be physically and emotionally torn to shreds – to be on this incredible cycle of health and sickness – and all the while feel responsible for another being, a child or family that is totally dependent on you. It's the same story with so many vets, it's too big a burden for them. You know the rug can go out from under you at any moment, you know how unstable you are, but there's the kid, and there's no one but you..."

He watches Zack, busily demonstrating for his friends the art of stacking wood.

"I've been lucky. Zack became the center of my life after my marriage broke up. For a few years there I pulled it together enough to work. By early '78, I was a partner in a couple of print shops in Lewiston. A customer who worked for the Idaho Conservation League talked me into doing copying at a reduced rate. I read some of the stuff I was copying for him. It was about herbicides, and it bothered me, though at the time I didn't connect it with my problems or with Agent Orange and Vietnam.

"I read in the stuff I was copying that five national forests in north Idaho had come out with a draft EIS on herbicide use on sixty thousand acres there, using ten herbicides, mostly aerial applications of 2,4-D and 2,4,5-T mixed. I saw the EIS, and maps of the areas they intended to spray. It was all my favorite places in Clearwater and Nez Perce – all my favorite fishing streams were set up to be plastered. That hit me right where it hurt. Those are the places my dad and I fished when I was growing up – White Sand Creek, where I discovered my being, where I rediscovered it after 'Nam.

"The linchpin was their risk assessment claiming that significant drift would not occur. I knew from dropping leaflets out of helicopters that that was just untrue. Forests aren't like farmland. You don't just fly the chopper over level ground. There isn't any. You have hills, swales, trees, updrafts, cross winds, steep slopes to contend with. You're turning constantly, and every time you turn, those rotors push air forward, to the sides, backward. The notion that all the herbicide would drop to the ground was a crock. Even in a no-wind situation, the helicopter makes its own wind.

"So my interest was aroused, to put it mildly. I got very involved with CATH. I started really reading that CATS information packet, and when I hit the research on industrial exposure cases, things started falling into place – those workers had the same fucking symptoms I had, and other guys like me, other vets. I made a list at one point of similar symptoms and came up with over a sixty-symptom match. That did it. I swapped assets for liabilities with my partner and got out of the printing business to join the lunatic fringe."[4]

"Those Chemicals from Vietnam"

In the years following their military service in Vietnam, countless veterans suffered problems similar to Paul Merrell's. Some have succumbed to cancer, others have fathered children with severe birth defects (one of whom was chosen national poster child for the March of Dimes)[1]. Most of them are afflicted with difficult emotional problems. Until 1977 and 1978, very few veterans connected their health problems with exposure to chemical weapons.

When the news of Agent Orange disabilities became known, many veterans, including Paul, reported their problems to the VA, claiming service-related disability. Both their problems and the VA's treatment of them were discouragingly consistent, as veteran after veteran testified at a Vietnam Veterans Against the War Investigation of Agent Orange in December 1979.

> JOHN LINDQUIST: My medical problems since my return from Vietnam have been this rash, and numbness that comes and goes in my hands. It wasn't too bad when I first got out, but now that I'm getting older, thirty one—real old! – I can't sleep on my side or my stomach. I have to sleep with my hands folded across my chest, or with my hands out away from my body, and it interrupts my sleep. Just recently a new development has been that I have four times the normal liver function count...
>
> Six of us went out to get tested at the VA, in April, '78. We were given a chest x-ray, blood test, and urinalysis, and were told that we'd be called back in two days. Two months later, I called again and said, "Are you going to call me back?"

They said, "Well, if we didn't call you back, that means you're all right." Well, little did they know that we had called the Disease Control Center in Atlanta and asked if Agent Orange poisoning could be tested in that way. They told us that we would have to have eaten it for breakfast for it to show up in a blood test! We busted the VA for their phony test, called the press, held demonstrations, and finally the VA has admitted they don't have a test...

The VA said they were gonna retest some of us, because of the squawk we made about the phony tests. A few of us were called in October, '78. The only thing different we received was to fill out the "magic" four-page form, which asked us a lot of questions on what time of day it was when you were exposed, what date, where you were, which way the wind was blowin', what did you feel like – a very bizarre form!

The VA's a training facility, veterans are generally guinea pigs, and it hampers the effectiveness of any testing the VA could do. I also have a problem with my short-term memory. I'm concerned about it, if I add it to the rest of the symptoms. They are mild, compared to other brothers sitting on this panel. I know that there's thousands of other veterans out there like me, and thousands who are in worse shape than the people in this room. And there's hundreds of veterans, if not thousands, who have already died from this.

FLOYD MOORE: When I got home and had problems I looked for help. They would refer me back to the VA hospital and the doctors would always say I had a dishidrosis that was common among black people – rashes, peeling, swelling hands, numbness, excessive tiredness, mental lapses where you would forget things that you wanted to say and couldn't say them ...every year it gets worse and I've been told the same thing over and over and over ...

American soldiers in a defoliated area in Vietnam

I won't take their medications because the medication is Thorazine and all those other big-name pills they give you to keep you drugged up, so you can't do anything, you can't fight them.

DENNIS KROLL: I was wounded in Vietnam, and did sustain some nerve damage in my left hand...The numbness spread, mainly in my right hand, the one that wasn't damaged. That numbness has now spread completely through my arms...I have had bouts with headaches that have lasted for two weeks at a time, bouts of nervousness that I can't pin on any stressful situations in my life, periods where I'm unable to relax at all, times where I can't concentrate, and seldom can pick up anything to read because I can't remember the lines I read previously.

...Back in 1972, I was given an electrical test I believe it's called an EMG [electromyocardiogram] where they took readings off my fingertips. Well, the amount of electricity they ran through my arms was enough to literally lift me right off the table, and they were still having trouble getting a reading! This was from both arms, not just the one that had been physically damaged...I requested a copy of that test, and all my medical records. I received two copies, and in neither one was there the results of that test I took in 1972...From the disability claim that I filed, I have received only a two-line correspondence from the Regional Office in Milwaukee: "Please verify exposure to toxic chemicals." Unless it's in my military or medical record, there's no way I can make that verification. That's what's been happening and the treatment I've been receiving from the VA.

BILL WARE: We were up near the DMZ working with the Green Berets...I heard a lot of choppers coming in and suddenly the area was fogged in, like a mist. The call came down the line that we should use our gas masks that we'd been carrying for about a year without using...The people involved started coming down with a lot of problems—watery eyes, persistent coughs, nervousness, skin rashes, itching... We were given medication. Some people had fevers; we were told that it was probably slight colds or jungle rot or "fever of unknown origin." Most of the people who had been seriously affected continued to have problems. I had a number of problems: a continual cold, running sinuses, a slight cough.

After leaving the service I worked for a couple of months at a foundry in Milwaukee—International Harvester. I ended up in the hospital for pneumonia. Since then I've been having episodes two, three times a year. I've been in and out of hospitals and under a private doctor's care. I've been diagnosed as having acute bronchitis, chronic asthma, sinutosis rhinitosis. I've had severe skin rashes. The VA has diagnosed me as paranoid schizophrenic with severe mental problems.

DAN MOELLER: My exposure to Agent Orange and other related herbicides was in 1972 toward the end of the war. Officially we had stopped spraying areas. However, I was pulled on a duty roster in Tan San Nhut and Bien Hoa to load trucks, helicopters, standard-wing aircraft, and even loaded the sprayers. It came in 55 gallon drums, and when I asked what the noxious petroleum smell was, the guy said, "Don't worry, kid, it's only weed killer."

The next day after the duty roster, I developed what was commonly known in Vietnam as "fever – unknown origin" and gastritis. I was sick for one day and recovered. About a week later, I developed a severe rash on my right thigh and some dotting rashes on my wrists. The dermatologist at Third Field said, "Here's some cream and ointment, here's some pills, forget about it. It's just one of those weird things you get in Vietnam."

Five years later, I woke up one morning with no feeling in my left hand. Gradually, it started to come back, like I had slept on it. Day after day the same feeling continued. It started into my right hand. Within the past eight months, hair has begun to fall out of my wrists and my right leg...

As for the VA – I had an EMG, which is an extremely painful test of electrical nerve conduction and studies both the nerves and the muscles, and how their patterns trace. This test came up grossly abnormal over the previous test which I'd had the year before. This test was conveniently lost by the VA hospital. To date, nothing has been found of this exam... The VA has continually denied that I have exposure.

LESTER HARRIS: Once I was in Vietnam I developed a rash between my legs. I went to the medical center and they told me it was a heat rash. I said O.K., I'll buy that – what else could I do? They gave me some cream and said, "It'll go away in a few days." So I used the cream and it didn't go away. I went back. They gave me some more cream and that was that; they said I'd get over it. But I didn't get over it because I still have that same rash now.

Sometimes, three times a day, this chopper would come by with his extended mosquito wings, dropping [the spray]; there was a mist falling and we'd see it coming and split for the hootch. It happened off and on every day. Then I started complaining of headaches. I went back to the medics and they just said it would go away...

I get back here. In 1971 I got married, my wife became pregnant and there was a child born – a baby girl with one kidney and one lung. She was born at five in the morning and died that night. I asked the doctor the cause of death and he said, "She died." That's what they told me. That's it, she died. I said, "What did she die from?" They said, "She only had one kidney and one lung." But what caused her to have one kidney and one lung? They said, "Well, heaven knows."

Before I knew she was dead I was riding up in the elevator. They asked me, "Mr. Harris, would you like to donate your baby's body to science?" I said, "My baby's body to science?" They said, "Well, your baby's dead, didn't they tell you?"

MARY A. WACHTENDONK: Jim now has many symptoms of dioxin poisoning. Since his release from the Army in 1971 he has been hospitalized for stomach and gastro-intestinal problems...He suffers from rashes on his abdomen and legs, numbness in his fingers and hands, loss of hair, headaches, severe fits of depression and paranoia, and a general feeling of not being able to fit in this society.

Our two children, a girl and a boy, both have brain disorders. Our daughter was born with hydrocephalus and underwent brain surgery at two weeks of age. She is under medication to control epileptic seizures and frequently suffers from high fevers...

Our son Zachary at birth appeared to be perfectly normal...Later, it was obvious that Zack was not developing at a normal rate. We also discovered that he could not focus – his eyes moved rapidly back and forth. The official diagnosis by an ophthalmologist was "congenital nystagmus." His optic nerve has never developed. Also, recent blood tests done by his pediatrician indicate an abnormally high count of amino acids which is common in cases of mental retardation...Genetic testing revealed no cause for these birth defects. There are no abnormalities on either side of our families.

PAUL RAY JENSEN: It wasn't until 1975 that I began experiencing symptoms of my exposure, and this came after an extreme weight loss. I went from 165 pounds to 125 pounds in a matter of three months. Shortly afterwards, I began to notice numbness in my fingers and toes. These were the first physical indications.

Psychological symptoms appeared shortly after I came home from 'Nam. I have records of my admission to the Battle Creek VA Hospital Psychiatric Ward. The diagnosis was paranoid schizophrenic and passive-aggressive personality. I had blackouts, was easily upset, and experienced fits of anger. I had a hard time controlling my temper and was very disoriented. This led to my incarceration in 1975 for armed robbery. I also went through two divorces from 1971 to 1975. The numbness in my hands has been diagnosed as Reynaud's phenomenon by several private doctors, however the VA won't even recognize that![2]

* * *

"We were America's best, the cream of the crop," Paul Merrell says. "By the time they got through with us in basic, I could run a mile with a full pack in six minutes, could run 25 miles in a day. They didn't send anybody over there who wasn't in top physical shape."

As he says, Paul is lucky. His problems began while he was in Vietnam, since he was there long enough for symptoms to develop while he was still in the service. (The VA does not automatically recognize a health problem as service-related unless it develops during or within one year of military service.) The VA refuses to acknowledge Agent Orange as a possible cause of his problems, but it has granted him total disability for "a nervous condition" and chronic colitis.

Other veterans have not been so fortunate. The symptoms of Agent Orange poisoning are difficult to diagnose, for several reasons. As studies of industrial workers exposed to similar compounds in factory accidents demonstrate, many of the most debilitating symptoms do not appear for many years after exposure: the mental and emotional disorders, cancers, peripheral neuropathy (numbness in fingers, hands, toes, and feet). Many of the symptoms of toxic exposure are readily attributable to other causes such as flu, hepatitis, anxiety, stress, colds, drugs, or other diseases, and people often do not relate varied problems to each other or to a common cause[3]. Perhaps the greatest barrier to diagnosis, however, has been the refusal of the VA or the medical community to weigh the possibility that toxic exposure is the cause.

Another barrier until recently was veterans' ignorance of toxicity, an unawareness perpetuated by the government and obscured by the nature of the war itself. Aside from the effects of chemical warfare on our own troops, the Vietnam War sent its GI survivors home with wounds beyond the capability of medicine to diagnose or cure. No other major military conflict has left its veterans with the crippled integrity and blasted self-respect that infected American GI's upon coming home. Understandably, the veteran attributed his irritability, insomnia, memory lapses, rages, headaches, depressions, and other symptoms to the overwhelming malaise created by the war and public disdain.

What is difficult to accept is that the American government, through the VA, to this day prefers to believe that it turned "America's finest" into unemployable mental wrecks by sending them into a senseless war, rather than admit that it might have poisoned them with its own chemical-warfare agents. The implications of such an admission, both to powerful industries and to the national conscience, have yet to be confronted.

By 1977, increasing numbers of the nation's 2.8 million Vietnam veterans suffered illnesses and disorders no one could explain or cure. Isolated from society by the nature of the war, they were generally isolated by the nature of their symptoms from one another as well. Many of them shunned society altogether, including the society of their fellow veterans, from whom they might have discovered that they were not alone with their problems.

"I used to think, maybe it's my mind trying to shut out everything. Maybe I've had too much for my mind to handle, maybe it is mental. But I don't want to believe that." James "Peewee" Fortenbury recalls, as quoted in the Uhl and Ensign book GI Guinea Pigs. "I know that my liver disease and my enlarged spleen isn't mental...I'm in the woods right now about this Agent Orange poisoning. I didn't even know anything about it until I read it in the paper; all I knew is that I have a liver disease, I've been very depressed.

Of course the depression could be caused by things that's happened to me in 'Nam. But is it causing sleeplessness, liver disease, enlarged spleen, high fever?"[4]

**

In 1977, Maude de Victor, a case worker with the VA in Chicago, had never heard of Agent Orange until a dying Vietnam veteran told her his cancer was caused by "those chemicals from Vietnam." After he died and the VA denied his widow survivor's benefits, Maude made some inquiries about "those chemicals" and began noticing a pattern of symptoms in the veterans' files that crossed her desk. She started asking her clients, "You been in Vietnam? Got any kind of rash? Have any children with deformities?" Frequently the response was, "Yeah, how'd you know?"[5]

Within a few months, Maude had accumulated 27 cases of this "new disability" and located thirty more at the VA hospital, all from the Chicago area. When she was ordered by her superiors to stop her investigation, Maude contacted WBBM TV news correspondent Bill Kurtis.[6] On March 23, 1978, the news about Agent Orange poisoning of U. S. Vietnam veterans broke in Chicago with an hour-long television documentary. The show featured interviews with veterans, victims of domestic use of the herbicides, research scientists such as Dr. Matthew Meselson of Harvard University and the environmentalist Barry Commoner, and Air Force Captain Alvin Young, who had supplied the information on dioxin toxicity that spurred Maude's original interest. (On camera, however, Captain Young suggested there was little evidence that Agent Orange could have caused the veterans' symptoms.)[7]

Veterans appearing on the show reported birth defects in their children and health effects ranging from cancer to numbness of the hands, heart conditions, chronic fatigue, and irritability. An American secretary who had worked in an Army logistics office in Vietnam – the only woman at that time who had filed a claim with the VA for Agent Orange poisoning – told of four pregnancies lost through miscarriages and spontaneous abortions. Bob McKusick of Globe, Arizona, told of his and his neighbors' experiences with Forest Service spraying, and the owners of a horse arena in Missouri told of losing 67 horses after oil contaminated with dioxin was sprayed in the arena to keep the dust down in 1971. Both the owners and two children playing in the arena continued to have health problems seven years later.

The disclaimers by Dow and government officials on the show were undermined by independent scientists who had long experience with toxic chemicals.

"Dioxin is distinguished by its incredible toxicity," Dr. Meselson said. "We don't know why it is so toxic, we only know that it is, and there is furthermore very important suggestive evidence that it is highly cumulative, each little bit causes damage that resides in the body. Low-level exposure may not produce obvious effects today or tomorrow, it may be ten years, it may be twenty years. Thus you may be exposed to the compound and suffer no ill effects but you may have damage done that will manifest itself years and years later. Now we are talking about the possibilities of cancer, the possibility of mutagenic effects."

Questioned about the possibility of 2,4,5-T and dioxin residues being stored in the body, Barry Commoner replied, "It may well...be found in soldiers who were exposed to dioxin in Vietnam which accumulated in their body fat with no symptoms...except for the immediate skin symptoms, and then let's say ten years later they become sick and lose weight. They would break down that fat, releasing the dioxin into the body, and then symptoms would appear."

(Two years later, fat biopsies of Vietnam veterans detected dangerous concentrations of TCDD in their body fat many years after their original exposure in Vietnam.)[8]

Dr. Irving Selikoff of Mt. Sinai Hospital in New York summed up the problems with diagnosing illness related to toxic-chemical exposure. "With dioxin we do see things fairly early on. For example, our colleagues in the laboratory have shown us that quite rapidly we find effects on blood cells. We find effects on the liver, we find effects on the enzymes. You find effects on the immune system, and these are unusual.

"At the same time, it is difficult to categorize these vague symptoms at the moment. When somebody gets an effect on memory, how do you quantitate that? What kind of statistics do you have when somebody can't remember where he left his car? Or he has to stop work because he does not remember which exit to take on the highway, or when his sleep is now badly disturbed or he can't concentrate, or where he is now irritable all the time. When his wife says I can't live with John any more and he says, 'I know it...but I can't help myself.'"

The television documentary triggered intense, local public interest in the effects of Agent Orange on Vietnam veterans. The word quickly spread nationwide, alerting veterans – many of whom had not recognized a service-related cause for their problems – that they were not alone in their afflictions.

"Other veterans groups, which principally concerned themselves with issues of amnesty, less than honorable discharges, vocational counseling, psychological readjustment, and education benefits, became deluged with inquiries from veterans concerning whether their health problems could be related to Agent Orange exposure," a Senate subcommittee reported[9].

Groups such as Citizen Soldier, Veterans Education Project of the American Civil Liberties Union, the National Association of Concerned Veterans (NACV), National Veterans Law Center, Vietnam Veterans Against the War, Vietnam Veterans of America, Coalition of Veterans, Swords to Plowshares, and Flower of the Dragon suddenly shared a common interest in Agent Orange with one another and with groups formed specifically to address that interest, such as CAVEAT, Vetline Hotline, and Agent Orange Victims International. The Agent Orange controversy provided a common focus for veterans' organizations that differed widely and often bitterly in philosophies, politics, and social status. Many of these groups met in St. Louis in mid-1978 to form the National Veterans Task Force on Agent Orange, a national coalition chaired initially by Ron de Young of CAVEAT, who had introduced Maude de Victor to Bill Kurtis of WBBM TV News.

As the accounts of numerous veterans demonstrate, their treatment by the VA has been consistently inadequate and degrading. Veterans attributing any physical condition to exposure to Agent Orange have been treated as mental patients, denied compensation for their disabilities, deprived of medical attention for clearly debilitating physical problems, and branded as paranoid schizophrenics, aggressive personalities, depressive types, and other labels guaranteed to make pariahs of them. Despite intense public, congressional, and veterans' pressure, the VA has devised Catch-22 policies on Agent Orange disability, effectively closing its doors against the veterans it was chartered to serve by demanding impossible proof of exposure and equally impossible proof that exposure is the cause of their problems. (The VA has relied on the same Catch-22 requirements in rejecting all but twelve of the 943 claims filed by veterans exposed to radiation as human guinea pigs in nuclear bomb tests.)[10]

"No group of servicemen who risked their lives for their country in more ways than we may ever have realized have ever been more rudely treated on their return," said New York Republican Congressman Norman Lent at hearings of the Oversight and Investigations Subcommittee on September 25, 1980. "Sadly, whole hosts of Vietnam veterans have lost all trust and confidence in the very federal agency which was established long ago, specifically to provide for their needs."[11]

Veterans who have turned in frustration to the courts to force the VA to take responsible action have run squarely into roadblock laws which exempt the military from the laws of the land. A Supreme Court decision in 1950 (Feres v. U.S.) denies military personnel the right to sue the government for admittedly negligent acts which occurred on active duty, thus denying veterans access to the judicial system. The manufacturers of Agent Orange may also be immune from legal action if the court finds that "A government contractor is not responsible for mistakes for which the government itself cannot be held responsible" – what attorney Lewis M. Milford of American University's National Veterans Law Center calls "a contract law version of the Nuremburg defense."[12]

In response to increasing pressure from veterans, the public, and Congress, the VA proposed a study of the health status of the 1,160 men who had participated in Operation Ranch Hand. (Operation Ranch Hand, the code name of the U.S. military program for defoliation in Vietnam, developed from earlier, tentative Operation Hades missions.) The proposed study was considered to be seriously flawed in design by the National Academy of

Sciences, which shared veterans' concern about the impartiality and credibility of an Air Force conducted study. Since the VA showed no willingness to soften its rigid demands for proof of exposure, the Ranch Hand study would be little use to the majority of veterans, who could not provide such proof[13].

At a public hearing before the Subcommittee on Oversight and Investigations on June 26, 1979, witnesses charged the VA with "failing to study the Agent Orange problem, withholding veterans' medical records, and inadequately examining veterans who claimed medical problems related to Agent Orange." In addition, they charged that the VA consistently carried out its decision-making in secret, refusing to allow veterans any participation in designing its studies of the Agent Orange problem or formulating policies. It continued to distribute circulars to physicians and personnel of VA hospitals asserting that "there is no positive evidence of deleterious effects on the health of individuals exposed to these herbicides which is of a permanent nature."

In December of that year, Congress passed P.L. 96-151, the Veterans Health Programs Extension and Improvement Act of 1979, making the VA responsible for an epidemiological study of Vietnam veterans. The act's legislative history emphasized that the study should address particularly the concerns of "Vietnam veterans who had no faith in the government's ability to study the Agent Orange issue objectively." The VA, however, deferred to the act only in contracting outside the agency for design of the study, retaining control over its performance.

The epidemiological study thus designed in response to P.L. 96-151 limited the study – as before – to short-term examination of Ranch Hand personnel. It did not take into account the chronic and extensive exposure of ground troops to herbicides in their own camps, on roads and rivers, in their food, water, and air, and in smoke from the burning of defoliated areas. Other readily identifiable groups not considered in the study were helicopter crews who sprayed Agent Orange (these were not part of Ranch Hand), helicopter support units flying alongside spray helicopters to provide protection, combat engineers who operated heavy equipment to clear defoliated areas and then burn the debris, and army personnel who mixed and handled Agent Orange[14].

The most heavily and repeatedly defoliated areas were along roadsides and rivers in Vietnam to ensure visibility for travel. Jon Furst, the chairman of the National Veterans Task Force on Agent Orange, has noticed a significant percentage of Agent Orange symptoms reported from troops involved with transportation, road building and maintenance, and other operations conducted extensively on or around roads and rivers[15]. Certain readily identifiable groups in that category could provide excellent material for a study.

The 116th Engineer Batallion of the National Guard, for example, was employed chiefly with improvement of Highway 20 in the central highlands, clearing and burning defoliated areas. These men all came from the north central Idaho area, thus sharing a similar background both environmentally and geographically, and most of them returned to the same area after the service. Many of them have reported serious health problems of the kinds associated with Agent Orange exposure. A study of all the men in the 116th would have been fruitful, but neither Citizens Against Toxic Herbicides nor former Idaho Senator Frank Church were able to pressure the Department of Defense into providing the names and addresses of the men in that battalion[16].

VA reliance on the Ranch Hand study, and insistence on rigid proof of exposure and causality, are inconsistent with VA's own regulations, which require that any reasonable doubt involving a veteran's claim be resolved in favor of the veteran[17]. Since herbicides were applied so heavily and ubiquitously wherever U.S. troops operated in Vietnam, it would be reasonable to presume exposure to be an inseparable element of Vietnam service. The finding of dioxin residues from the herbicides in shellfish off the coast of Vietnam – indicating how widely the herbicides were dispersed in the environment and food supply there – strongly supports such a presumption[18].

The similarity between veterans' health problems and the disorders suffered by victims of domestic and industrial exposure suggests a reasonable probability that their problems are related to their exposure in Vietnam, complicated perhaps by re-exposures at home[19].

In defiance of such suggestive evidence, the VA remains implacable. The decisions of its medical review board are final, and are not subject to review by the courts. In other words, veterans cannot sue the VA for the treatment they deserve[20].

Vietnam veterans, Jon Furst testified, are "experiencing a crisis in confidence regarding their governmental institutions. I would remind you that these Americans have worked with the government before...and that many perceive each contact to have resulted in disaster, betrayal, and/or denial. It is, perhaps, one aspect of reality for many veterans that their military experience has resulted in a very peculiar relationship to their government."[21]

In December 1979, an Interagency Work Group was created by President Carter in response to growing concern about herbicides and dioxin. It was charged with responsibility to "oversee, coordinate, and set priorities among the federal government's research activities designed to relate exposure to phenoxy herbicides to long-term health effects." The Work Group consisted of the Department of Health and Human Services (HHS), the Department of Defense (DoD), and the VA, with observers from the EPA, the USDA, the Department of Labor, the White House Office of Science and Technology, and the Office of Technology Assessment.

The Work Group, however, proved to be little improvement over the VA itself in examining the Agent Orange question. It persisted in relying on a Ranch Hand study for determining V. A. policy, and relied on unverified evaluations of the military's exposure records, in spite of a General Accounting Office report that the DoD had been "inaccurate in its reports on ground troop locations and likelihood of exposure." Although the Work Group made some laudable suggestions for expanding the epidemiological study mandated by Congress in P.L. 96-151, the VA showed no signs of any willingness to change its policies should an epidemiological study demonstrate a reasonable likelihood of adverse health effects from Agent Orange. The VA persisted in requiring scientific proof of a "definite linkage between exposure to herbicides and a specific liability"—proof that no epidemiological study, no matter how well designed, could ever provide.

The Subcommittee on Oversight and Investigations found that the Work Group failed to review or improve the VA's Agent Orange screening examination, and that "the information disseminated by the VA has prolonged the use of incomplete and often useless examinations."

"The VA continues to distribute incomplete and often misleading information to VA facilities and veterans themselves," the Subcommittee said. The role each participant – particularly the VA and DoD – should play in the decision making process and establishment of policy was never clear, and by 1981 it was apparent that VA policy remained immune to recommendations from the Work Group, Congress, the veterans, the public, or anyone else[22].

In contrast to the VA's wholesale rejection of Agent Orange claims, individual states have shown an active and compassionate interest in their Vietnam veterans' problems. In December 1981, for example, the Texas Veterans Agent Orange Program, administered by the Texas Department of Health, proposed six pilot studies developed by the faculty of the University of Texas. The proposed statewide studies included a Vietnam Veterans Mortality Study (comparing causes of death among Vietnam veterans with causes of death in veterans who did not serve in Vietnam); an evaluation of birth defects in offspring of fathers exposed to TCDD; cytogenetic testing (for chromosome damage); testing for sperm abnormalities; immune evaluation of veterans exposed to Agent Orange; and fat tissue analysis for TCDD.

The Texas studies are necessarily limited by their sample population size, uncertainties regarding exposure, and difficulties presented by confounding factors, such as domestic use of the same chemicals. The study protocols also recognized the limited utility of the studies to the veterans involved: Individuals whose test results are positive cannot be offered therapeutic manipulation or corrective intervention. There is no known way of reversing chromosome damage or sperm abnormalities." But a reliable study showing a link between veterans' problems and their exposure to Agent Orange would provide scientific credibility to veterans' claims and would give Congress a sound basis for legislation forcing the VA to change its policies.[23]

In 1980, the staff of the Subcommittee on Oversight and Investigations prepared an excellent report on the history of the Agent Orange issue, VA policy, legislative actions, and veteran involvement. The report concluded that with certain changes – notably extensive public participation and establishment of the Work Group in a leadership role – the Work Group represented the best hope for development of a satisfactory solution to the Agent Orange issue. The report made concrete recommendations for VA policy that included an outreach program to locate Agent Orange veterans, and the establishment of an "Agent Orange protocol" for identifying and treating Agent Orange symptoms[24].

Unfortunately, the change in administration following the November 1980 elections prevented this fine staff report from being completed or published. In June 1981, the U.S. House of Representatives approved by a 388-0 vote legislation directing the V.A. to provide hospital care and priority medical attention to Vietnam veterans whose health problems may be linked to exposure to Agent Orange. Determination of what health problems should be recognized as Agent Orange-related, however, would be up to the discretion of the VA and its individual physicians. A similar Senate bill would require the chief medical director of the VA to determine the nature of Agent Orange symptoms, a roundabout way to force the VA to establish rules.

Such congressional action is a step in the right direction, though it falls short of the subcommittee's and veterans' reasonable demands. Representative Thomas A. Daschle, a Vietnam veteran, told the House the legislation "will tell the Vietnam veteran, 'Yes, we're listening.'"[25]

More than "listening" is in order, however. "Before Vietnam deflated a robust tradition of Yankee patriotism, military service was an honorable calling," Uhl and Ensign wrote in *GI Guinea Pigs*[26]. The deteriorating bodies, hearts, and minds of Vietnam veterans are a monument to governmental hypocrisy and disregard for those who once proudly served their country – hardly an inspiration to the next generation of draft-age young men.

"Even if we never solve the mystery completely, the important thing is to stop telling the veterans there's nothing wrong with them," says Dr. Ronald Codario. "One hundred fifty of the two hundred people I've examined have porphyrin problems, numb fingers and hands, and an unknown molecule coursing through their systems. It's time to quit the denials and help."[27]

* * *

At the heart of the controversy over herbicides are basic moral and ethical questions long shunned by manufacturers of pesticide poisons. However, chemical companies are owned by stockholders whose moral and ethical values may, in some instances, take priority over economic considerations.

In 1973, shareholders in Distillers Company Biochemicals Limited, marketers of thalidomide in Great Britain, in effect overturned a meager court-approved settlement when they pressured the company into establishing a $47 million trust fund to provide for thalidomide victims[28].

In recent years, the traditional champions of morality and ethics – the churches – have recognized the serious information gap between the average stockholder and the companies he or she partly owns. As corporate investors themselves, churches and religious institutions have recognized a unique opportunity to serve as interlocutors between stockholders and corporate policy-makers, adding a much needed perspective on social responsibility to economic decision-making[29].

The Interfaith Center on Corporate Responsibility (ICCR), sponsored by the National Council of Churches, is an organization of church and religious investors who are concerned about the social and environmental impact of corporate behavior and the application of social criteria to investments. Its members include representatives of seventeen Protestant denominations and more than 170 Catholic communities. Assisted and coordinated by ICCR staff and work groups, members initiate and participate in such activities as public hearings, testimony before government and other agencies, extensive research, open letters, litigation, publishing, and boycotts (e.g., against J.P. Stevens for its labor practices, or Nestle's for marketing of infant formulas in developing countries).

One of ICCR's most effective ways of introducing moral and ethical issues into corporate activities is through shareholder resolutions. Under the regulations of the Securities and Exchange Commission, any shareholder has the right to submit proposals on certain policies and practices of a corporation for consideration by all shareholders at the company's annual meeting. Church investors have exercised this right in order to raise questions about social issues, requesting information or asking companies to take specific actions. (ICCR itself does not own stock and rarely acts in its own name, functioning chiefly as a coordinating and information-gathering body and communications network.)

In both 1980 and 1981, several church and religious institutional investors presented a stockholders' proposal to the annual meeting of the Dow

Chemical Company, manufacturers of 2,4-D, 2,4,5-T, and Agent Orange. The 1981 proposal asked Dow stockholders to support a resolution directing Dow to establish an independent review committee to evaluate "the existing and potential health consequences of 2,4,5-T, silvex, and their derivatives, and make recommendations to the Board relating to the justification of continued production of these herbicides."

The resolution cited the 1979 EPA emergency suspension of 2,4,5-T and silvex, and two 1980 National Cancer Institute studies finding dioxin to be carcinogenic. It noted that Dow's expenses for defending 2,4,5-T exceeded its profits from the herbicide, which accounted for less than 0.2 of the company's annual sales. (The basic principles of the stockholders' resolution – including its call for a moratorium on international production of 2,4,5-T and silvex – are expressed in the National Council of Churches' Resolution, Appendix E.) Dow's board of directors issued a statement in opposition to the resolution, urging stockholders to vote against it. Dow's essential argument was that the EPA hearings on cancellation of 2,4,5-T and silvex served the same purpose as the proposed Dow review committee, making such a review committee superfluous. (This was a more revealing argument than shareholders could have known. By the time of the annual meeting on May 8, 1981, Dow had already maneuvered EPA into secret negotiations to settle the issue without proceeding with the hearings.)

The board of directors' statement insinuated that the proponents of the resolution relied solely on one principal scientific adviser, Dr. Samuel Epstein, professor of occupational and environmental medicine at the University of Illinois School of Public Health. Dr. Epstein, a human pathologist and toxicologist and author of The Politics of Cancer[30], had presented a statement in support of the shareholders' resolution in 1980, reviewing a wide range of governmental and scientific studies confirming the hazards of 2,4,5-T and related chemicals (see Appendix B). Contrary to Dow's insinuation, the research Dr. Epstein summarized was not his own, but had been conducted by reputable scientists and institutions throughout the world.

The 1981 annual meeting was addressed by Dr. Marvin S. Legator, the director of the division of genetic toxicology of the University of Texas medical school, and a former consultant with Dow Chemical Company. He supported the resolution with a review of Dow's own policies for monitoring carcinogenic or mutagenic effects of its products on its own employees, and presented summaries of two recent studies from Sweden and New Zealand implicating 2,4,5-T as a cause of cancer and birth defects, respectively (see Appendix C). The Dow statement claimed that the National Cancer Institute studies cited in the resolution actually supported Dow's position on 2,4,5-T, and referred extensively to a British government report concluding that 2,4,5-T herbicides were safe. (Dow did not mention that on the same day the British study was released, the twelve-million member Trade Union Congress of Britain unanimously endorsed an embargo of 2,4,5-T. The British trade unions had released a rebuttal of the government report, asserting that the report downplayed convincing evidence against 2,4,5-T, that it employed arbitrary criteria for establishing the "safety" of 2,4,5-T, and that it sought no input from key parties in the complaint against the herbicide – the workers who were subject to exposure.)

The stockholders' resolution brought by the churches was defeated two years in a row by Dow stockholders, but as Reverend John Jordan of the National Council of Churches points out, it achieved a larger purpose of bringing the issue to public awareness and educating both stockholders and the public.

"As chairperson of the Vietnam Generation Ministries Working Group of the National Council of Churches," Mr. Jordan said, "And as its representative on the National Veterans Task Force on Agent Orange, I have personal contact with many Vietnam veterans who have serious health problems which they attribute to exposure to Agent Orange. We know that over thirteen thousand Vietnam veterans in the State of Minnesota alone have asked for medical examinations or other help to diagnose and cure the unprecedented and otherwise unexplained common symptoms they have, including various combinations of cancers, skin and nervous disorders, miscarriages by their wives, and birth defects in their children...The further study called for in this resolution is the least we should do as an example of ethical and moral constraint in our corporate social responsibility for the possible health hazards of this product."

Mr. Jordan is a staff member of the Board of Global Ministries of the United Methodist Church, based in New York City. For the last fifteen years, he has worked extensively on international issues and international student programs. His work with the Vietnam Generation Ministries Working Group of the National Council of Churches has centered on issues such as jobs, education, psychological readjustment, and other general health problems of Vietnam veterans. The Working Group has devoted particular attention to incarcerated veterans and to veterans and their families who are experiencing special health problems attributed to exposure to Agent Orange used in Vietnam. Mr. Jordan first was alerted to these problems by a veteran, Joe Bangert, who had organized a self-help group of veterans in Massachusetts.

"It was a new subject for me. I was not at all acquainted with the history of Agent Orange and with herbicide effects," Mr. Jordan says. "You have to understand that when I say Agent Orange, I mean the specific herbicides – 2,4-D, 2,4,5-T, and dioxin – used extensively in Vietnam and the U.S. Although my involvement has been primarily with the veterans, my concern includes the people who live in the forests and along sprayed roadsides and agricultural areas as well as workers exposed in their workplaces, who all share the problems the veterans have.

"Anyway, it was a new subject, and I did my best to find out more about it. I did a lot of reading, and began coordinating with other church people who became involved."

Together with Louise Ransom, a founder of Gold Star Mothers for Amnesty after she lost her son in Vietnam, Mr. Jordan worked with the National Association of Concerned Veterans to organize a special gathering of Agent Orange activists at their annual meeting in 1979. Through this gathering, Mr. Jordan became involved with representatives of local groups of Agent Orange veterans who came together to form the National Veterans Task Force on Agent Orange.

"The immediate problem here is the health of the veterans and their families," Mr. Jordan says. "That includes people in the broader group – on roadsides, in the forest, in the workplace. It's more than a matter of medical attention, because the greatest problem is getting officials or the public to acknowledge that a problem exists. It's a matter of political awareness and action, not only scientific education. Science will not answer all the questions – at least not in the near future – and depending on corporations or government to take responsibility won't work. There has to be public action.

"The long-term problem involves more than Agent Orange. It involves the attitudes of people – individuals and society – toward the world. I mean the environmental and ecological problems facing the whole earth. Changing corporate behavior – making corporations accountable to wider ethical and moral considerations – is an enormous thing to hope for. It essentially means changing the philosophy of our society, but there are times when a change of that order is necessary for survival.

"The complexity of the issue can be dealt with by simplifying the way information is presented. Knowledge must not only be made available to people, it must be available in language they can understand. There's a term we use in various groups I work with – demystify. We must work to remove the mystique of information as the property of experts – whether scientific, economic, political, or ethical. If you make information available and comprehensible to everyone, you reduce their dependence on experts and give them the means to become directly involved in decision-making. Demystifying the processes of society allows people to be more involved at all levels – this is real democracy.

"It seems to me the founders of our nation established the political principle in the basic rights to 'life, liberty and the pursuit of happiness.' All three are important, but there's a good reason why life comes first. You can't have the other two without life. It has priority over all. It's a fundamental biological and religious value, not just a political slogan."

Church leaders such as Mr. Jordan, working together through NCC and other national and international organizations, represent a growing universality of concern within traditional Christianity that extends beyond sectarian boundaries and rigid ecclesiastical limits.

"The importance of the issue – of Agent Orange and toxic chemicals – transcends any particular group," John Jordan says. "Compassion—human or divine—does not stop with the 'saving' of a soul.

"We didn't come here to bury dead babies," says a priest in Peru.[31] Human suffering and the threats posed by nuclear and chemical technology have involved the church in broader issues of social justice and environmental concerns, leading it inevitably into realms of politics and economics where angels traditionally fear to tread. There is nothing new in this ecumenical emphasis. It has its roots in the teachings of sages – Saint Francis and Saint John of the Cross, for example – as well as in Jewish and Christian scripture. The nuclear age has changed only its scope and its explicit direction.

"We are living in times that have no precedent," Simone Weil wrote in 1942, "And in our present situation universality...has to permeate our language and the whole of our way of life. Today it is not nearly enough merely to be a saint..."[32]

* * *

John Jordan puts it another way. "If we care about persons anywhere, we must care about and care for the whole household – the whole human family living together in a beautiful, fragile world."

For Future Reference: The Alsea Record

... And the next day the women knew.
I don't know how they knew,
But they smashed every government in the world
Like a heap of broken china, within two days ...
Well, we had a long run. That's something
At first they thought
There might be a nation somewhere—a savage tribe.
But we were all in it, even the Eskimos,
And we keep the toys in the stores, and the colored books,
And people marry and plan and the rest of it,
But you see, there aren't any children. They aren't born.

—Stephen Vincent Benet (1938)

The Five Rivers valley where my children were sprayed with herbicides in 1975 lies in the Siuslaw Forest in the coast range of western Oregon, midway between Alsea and Waldport. It follows the course of the river called Five Rivers, which runs roughly northward to empty into the Alsea River. The valley floor is narrow, varying from 500 feet to a half mile wide. The sides of the valley are steep and heavily forested, broken in places by flood plains and fertile bottom land of the major tributaries which give Five Rivers its name: Lobster Creek, Cascade Creek, Buck Creek, Crab Creek, and Green River. Lobster and Crab creeks were named for the huge populations of unusually large crayfish that supplied a small commercial fishery in years past[1].

Eighty years ago, this little valley represented a microcosm of Oregon coastal mountain life. It was more heavily populated then, with homesteads on most of the level land. Most of these are gone now, and the forest has moved in to reclaim them. You come across a rotten corner post, a rusting donkey engine, an overgrown mill landing, or a bed of daffodils blooming incongruously in some desolate clearing far from any road, and the forest seems almost a kindness. It will endure, it will survive our mistakes.

Until the end of 1979, a local mill employed most of the workforce of the valley. Since it closed, many families have left. About fifty families remain in Five Rivers valley, many of them recent arrivals of the last ten or fifteen years. Families that leased mill land have either moved away or are preparing to. Others own small acreages, which they farm to provide varying degrees of self-sufficiency. The dwindling elderly population clings to what has been home for too long to contemplate any other. The younger, growing families settled here and stay for reasons as varied as their histories.

In 1972, Debbie Marano and her husband moved to Five Rivers, John to work at the local mill and Debbie to raise the family they both planned and hoped for.

Debbie had grown up in Oregon, spending her childhood first near Bend, at the Fall River Fish Hatchery on the Deschutes River, and then in Alsea, at a fish hatchery adjacent to Bureau of Land Management (BLM) land. It is possible that she was first exposed to forest herbicides during those years, but she was unaware of it at the time.

In May, 1975, three years after moving to Five Rivers, Debbie spontaneously aborted her first pregnancy after fourteen weeks of gestation. Spontaneous abortions, or miscarriages, in the first few months of pregnancy are not uncommon, and many pregnancies are so terminated even before the woman knows she is pregnant. Doctors, unable to explain why this occurs, often encourage a woman to become pregnant again, assuring her that the miscarriage is nature's way of disposing of a defective embryo.

"Involuntary" abortion, however, would be a better term than "spontaneous" abortion. It is well known to medical science that maternal or paternal exposure to radioactive elements, certain drugs, pesticides, and other chemicals can cause death or untimely expulsion of a fetus or embryo, either by directly affecting the embryo, by causing the mother's uterus to hemorrhage, or by altering the transfer of vital nutrients across the placenta to the developing child[2]. Some of these substances can affect the genetic structure and development of either the father's sperm or the mother's ovum prior to conception, altering embryonic development or metabolism in such a way that survival is impossible and abortion occurs[3].

By the late 1960s, animal experiments had demonstrated that phenoxy herbicides (both 2,4-D and 2,4,5-T) could affect reproductive success when administered to female animals at critical stages of pregnancy, usually in relatively large doses[4]. Few if any studies were done to determine the effects of chronic low doses administered prior to conception to either the father or mother animals.

Determining the effects of toxic substances on human reproduction is complicated by many factors. Experimental animals often do not respond in exactly the same way as humans to toxic agents (rats, for example, do not miscarry or abort, but reabsorb their embryos or fetuses)[5]. Women frequently miscarry an embryo at the time a menstrual period is due, thus never knowing they were pregnant. When a known miscarriage occurs, the mother will seldom connect the event with exposure to chemicals – especially if neither she nor the father were aware that exposure occurred. Humans are exposed to so many different toxic chemicals in their environment and food supply that it is almost impossible to single out any particular chemical – or combination of chemicals – that may be affecting reproduction. Epidemiological studies of specific groups of humans (those exposed to radiation, for example, or living or working under conditions that expose them to more of a particular substance than the rest of the population) can only suggest the probability that human reproduction is being affected.

Debbie and John knew nothing about herbicides when they attended the first meeting of Five Rivers residents to discuss Forest Service spraying in early 1976. Nothing about the first miscarriage the year before had seemed to indicate any unusual problem, and they were hopeful about Debbie's second pregnancy. At the meeting, they learned of the possibility that herbicides could cause miscarriages and deformities, and when Debbie's second pregnancy ended in a spontaneous abortion several months later in April, she was concerned enough to submit an affidavit in support of CATS' federal suit for an injunction against further spraying in the Siuslaw. Six months later, in October 1976, her third pregnancy aborted at eight-weeks gestation.

Debbie's doctor had her thyroid checked in December 1976, and found it normal. Two months later, in February 1977, she lost her fourth child at twelve-weeks gestation. At this point, her doctor also began to take an interest in the herbicide question. He contacted scientists at Oregon State University (OSU) about the possibility of a connection between Debbie's inability to carry a child full term and the widespread use of herbicides in her neighborhood. He was referred to EPA.

In March 1977, Debbie and her doctor met with Dr. Michael Watson, a toxicologist with EPA, who took blood samples from both Debbie and her husband for chromosome analysis. The same day, Watson came to Five Rivers and took samples of deer, elk, and beef from the Maranos' freezer and a sample of their garden soil. Two months later, in May, Debbie's physician performed hysterosalpinography (x-ray examination of the uterus and fallopian tubes) and informed Debbie that her child-bearing organs were healthy and normal.

In August, after repeated requests for results of the tests EPA was running on the samples taken in March, Debbie received a letter from Watson stating that both the Maranos' chromosomes were normal and that no dioxin (TCDD) had been found in any of the samples he had taken. "It is my definite opinion," he wrote, "that TCDD involvement in your inability to reach full term is extremely remote if not totally impossible."[6]

Two months later, however, in a telephone conversation with biologist John Noell, Watson stated that a second testing of the Maranos' soil sample had shown 98 parts per trillion TCDD. A reporter from the Eugene Register-Guard who was investigating the matter also called Watson and was told that the test showed 92 ppt TCDD, which Watson said was not significant and could not have caused Debbie's problems. Debbie learned of the positive soil sample only through Dr. Noell and the Eugene reporter. (EPA later denied ever having found dioxin in her garden soil.) Debbie lost her enthusiasm for gardening and started buying vegetables from the market in town. In

1978, she gave birth to the couple's first and only child, a healthy boy named Matthew.

"I was told when I was pregnant with Matthew that if I had another miscarriage, the E.P.A. would like to take the fetus for analysis," Debbie says. "After he was born they came and took a milk sample from me. I always wanted a career in science, and now I have it – a guinea pig[7]. But what good will that do for me or the babies I might never have? It seems to me they're doing all this backward. Why didn't they find out all this before they ever sprayed?"

In March 1975, the same spring that Debbie suffered her first miscarriage, BLM sprayed its lands in the Alsea area with what some people reported was an unusually high concentration of herbicides. Within the ensuing weeks, a number of women in the immediate area also miscarried. Three years later, Debbie joined those women in writing a letter about the miscarriages which alerted EPA and the nation to the fact that, as one woman said, the herbicides "are killing our babies."

Debbie and her family moved back to Alsea in 1979 after the mill in Five Rivers was sold. Valley residents still wonder about the land the Maranos rented in Five Rivers. Debbie's next-door neighbor died of cancer in 1978, and in the other house on the property, a young man who worked at the mill was told he had terminal cancer in 1979. (He died in 1981.) The woman who rented the Maranos' house after they moved miscarried in early summer of 1979, within weeks of a miscarriage suffered by the woman next door.

Debbie and John were hopeful that in moving to Alsea they would be exposed to fewer herbicide applications. In 1980, Debbie became pregnant again. In the fall, a Christmas-tree farm up the hill from their new home was sprayed. Soon afterward, both Debbie's cats aborted their litters, and three weeks later, Debbie miscarried again. The aborted fetus was severely deformed. Within the next year, Debbie miscarried again, shortly after neighboring land was sprayed. Debbie could not find out what had been sprayed on the Christmas-tree farm, but her doctor is now convinced that the herbicides are responsible for her difficulties in bearing children.

"Sometimes I think of moving again, but it's such a hard thing," Debbie says. "Anywhere there are farms and forests – anywhere John could find work

– they are going to be spraying. When I get discouraged and talk about moving again, he says, 'Where can you go – the Sahara?'

"But Christmas trees! Why spray Christmas trees with poisons? I don't understand."

* * *

Irene Durbin, her husband Lyle, and their three children lived in Five Rivers for seven years, from January 1972 until May 1979. Lyle worked for Brand S Corporation, which operated the local mill. The Durbins lived in the "mill camp," a small cluster of mobile homes located on Brand S land a mile from the mill. The company leased home sites to its workers and maintained a water supply, septic systems, power lines, and a television cable. When the mill and property were sold to Northside Industries in 1979, Lyle went to work for Brand S at another of their mills in Corvallis and relocated his family in Dallas, Oregon, where they now live.

In March of 1980, Lyle suffered two massive heart attacks and was unable to work for several months. Shortly after he returned to work, the Brand S mill in Corvallis was burnt to the ground in an accident. He has since been able to find work only at a local gas station.

"You can see why I still feel homesick here," Irene says, pointing to the flat fields outside the window of their mobile home. "What's hardest is to go out for a whole day and not see a single person you know, not one person to smile at or talk to. I miss that. I miss the feeling that even people you don't know very well are your neighbors, you can count on them in some way. Five Rivers was such a special place that way. I always loved it there. My dad worked as mill carpenter there at Brand S, and he and Mom had a house there.

"We were living in Salem then, and we used to go visit my folks, and sit on their porch and look at the stars and smell the air, and it always felt so good, just being there. But to think of being able to live there was just an impossible dream. When Lyle got a job there at the mill in '72, it was a dream come true. It was such a perfect place to raise your children and live your life.

"We didn't know anything about the sprays then. I don't think anybody did, really. Oh, they would notice the helicopters, or the spray trucks, but they were part of the work of the forest. You wouldn't pay them any more mind than a yarder or a log truck. I do know that the Forest Service sprayed every year during the time we lived in the valley, both from helicopters and along the roads with tank trucks. But if we noticed it, we didn't think anything of it. I don't remember whether we actually knew what they were spraying, even. "If we had thought about it, or known anything about the sprays, we would've become alarmed a lot sooner, I'm sure. All the water supplies in that valley come from small creeks and springs that drain the ridges and empty into the river. Our water came from Summers Creek – our drinking water and water for the garden. Summers Creek drains a huge area of Forest Service and private timber land, all the way up the ridge that divides Five Rivers from Lobster Valley. We bought our milk from a neighbor's farm and ate venison and other game from the forest. So there were many ways we could've been exposed to the sprays.

"But one thing for sure, we were never notified of the spraying. In the spring of 1973, in May, I had a miscarriage. I was two or three months pregnant at the time and didn't go to the doctor. It's 61 miles from the mill camp to the nearest doctor, so unless something's really serious, you don't go. A miscarriage that early in pregnancy isn't usually serious, just kind of depressing. The next year, in March or April, I had another miscarriage, also after two or three months of pregnancy. And soon after that, I became pregnant again. "That pregnancy was placenta previa. The placenta was located below the baby, collapsed on the cervix instead of above the baby where it's supposed to be. I went into labor five weeks early, and was hemorrhaging badly. They took me to the University of Oregon Medical Center, and my youngest son was delivered by caesarian section there. He was born with hyaline membrane disease in his lungs – the membrane isn't developed and doesn't allow the lungs to expand."

Outside the kitchen window, Irene's son is playing in a plastic wading pool with some other neighborhood children, as she talks.

"His lungs collapsed and he had an operation to insert a tube into his chest to draw the air out. His liver was underdeveloped and he had to have a complete blood transfusion. He was in the hospital for almost a month. I was there, too, for over two weeks.

"After he came home, he had constant respiratory problems over the next two years. He must've had pneumonia three or four times. But he wasn't the only one – our two older children had pneumonia, too. All of them had repeated nosebleeds, and we all kept having sinus attacks, headaches, and rashes-eczema. My husband had this kind of chronic diarrhea, which had never happened before."

Irene begins to prepare lunch for her youngster.

"Then in the beginning of 1976 the people of the valley got together at the old school house to talk about the sprays.

"We went to the meeting, and heard about all the effects the sprays could have on animals and people, about the birth defects and all the birth problems. And I have to say we were flabbergasted! We suddenly realized that all these problems had never happened to us before we moved here…

"We went to visit my folks that evening after the meeting, and told them about it and what we thought. They were like a lot of people around here, they'd noticed the spraying and assumed it was for mosquitoes and bugs, that it was all for our own good. I remember Mom saying, 'But they have to spray, or we'd be plagued with bugs!' Most of us had never heard of sprays for killing plants.

"My parents only came to a very slow awareness of what the sprays really were. Now they're militant! They're just outraged at the whole idea the government could even do such a thing. At the time, after that meeting, we were so emotional and upset, and so sure the sprays had caused our problems."

Irene calls her son in to lunch. He stands in the doorway in his underwear, dripping and woeful. Irene suggests that since he's too wet to come in, he might take his lunch outside. He grins happily and is gone with a sandwich and a plate of sliced tomatoes.

"It's really amazing to me that none of us knew anything before that meeting, and that so suddenly everyone knew. The news about herbicides spread like wildfire, and the next thing we knew, people from Five Rivers and Deadwood had organized and were going to court. We didn't feel we were the kind of people who could do anything like that – Lyle had his job at the

Images by Carol Van Strum
  EPA workers collecting samples at Five Rivers mill camp water supply (August 28. 1984)

mill, and I had the house and kids – but we were glad there were people who were doing something about the situation. And they did do something. They kept the Forest Service from spraying for a year, and woke up a lot of people who had never known about the herbicides before.

"Then, during the spring of 1978, the road alongside Summers Creek – our water supply – was heavily sprayed. After the spraying, they posted warning signs, but the only sign we noticed was far from our house, and it wasn't much of a sign, hardly bigger than a three-by-five card, and all it said was something about if you wanted to pick berries on this land you should contact the Forest Service for advice about an alternative place to pick.

"Those signs are so ridiculous, you wonder why they bother putting them up at all. They never notified us about that spraying, or cautioned us not to drink the water. Immediately after that, we could taste and smell something peculiar in the water. Two other families nearby, who also drew their water from Summers Creek, also noticed it. The taste and smell stayed in the water for several weeks after the spraying.

If dangerous chemical wastes have been hauled through this area, read these precautions carefully, especially if you may have been exposed to DIOXIN.

IN CASE OF **DIOXIN CONTAMINATION**

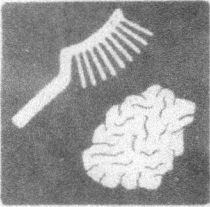
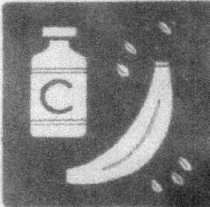

1. Wash thoroughly. Rinse eyes, ears and nose with sterile water. Scrub body with wire brush or steel wool.

2. Take vitamin C; eat fresh fruits; save your apricot pits.

3. Stay away from swimming and camping. Use bottled water if necessary brush...

4. Destroy contaminated pets and dispose of safely (six inches of concrete on all sides)

5. Avoid releasing dioxin molecules into the air in your home or oth...

6.

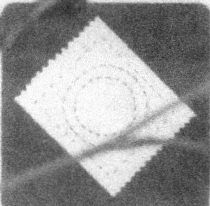
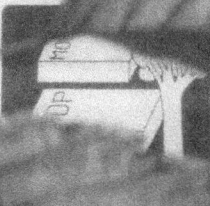

7. Use a condom for intimate contact with an uncontaminated other.

9. Remove plant away...

Disclaimer: There is no need to fear death from cancer after exposure... not survive. This geological area is...

Image by Carol Van Strum

Sign left at Five Rivers mill camp water supply after EPA's 1979 sampling

"That same summer Dr. Eldon Savage of Colorado State University's EPA investigation team visited us. Dr. Savage interviewed us and took water and sediment samples from the creek at our water intake. We heard nothing about those samples until the following spring. In March of 1979, the Lane County Health Department sent us a letter. It said that Dr. Savage had notified the county that dioxin had been found in our creek sediment at levels between ten and twenty parts per trillion.

"The day after that letter came, Dr. Savage telephoned me from Colorado. He said that he didn't feel the water was safe to drink, and that he himself would not drink the water. But we had been drinking it all these years! We didn't know how or when the dioxin got into it, whether it was from that roadside spraying the year before, or from other roadside sprayings in previous years, or from aerial spraying of the countryside over many years.

"None of the other families who drew their water from Summers Creek were informed of those test results. I called the Lane County Health Department and asked them to notify the others. All they did was send people a copy of their letter to me, but the letter said nothing about the danger of drinking the water. One lady thought the letter had been delivered to her house by mistake and brought it to me – it was just a photocopy of the letter to me, with my name still on it.

"When EPA was preparing their hearings on registration of 2,4,5-T, they had people checking on areas that had been sprayed aerially in this vicinity over the past few years. Between the years 1972 and 1976, 1,204 acres were sprayed with 2,4,5-T within a two-mile radius of Summers Creek. And if you look at the map, you can see that some of the sections sprayed in those years actually come within a quarter mile of our house. Every one of those years, spraying was done in that two-mile radius.

"Most of the spraying was done in the spring, and both of my miscarriages occurred in the spring. My son was conceived in the spring as well. The researchers were unable to find any roadside spraying information. No records had been kept."

* * *

No one may ever know what was sprayed along the roadside by Summers Creek in 1978. Tank trucks hosing roadsides do not have the dramatic impact of helicopters spreading poisonous fogs over the landscape, and most public controversy over herbicides has focused on aerial spraying. But for travelers on sprayed roads or people who live along them, roadside spraying poses an immediate, serious threat.

The State of Oregon, local county road departments, and federal agencies, such as the Forest Service, spray the verges of nearly every highway, road, and logging road in the state every year. Few records are kept of what is sprayed where, or when, or how much.

The crews hired to perform the spraying may be recruited from the local employment office, and in any case are often inexperienced. They are ill-advised of the hazardous nature of the sprays, although studies completed years ago showed chromosome damage in spray crews.[8] They commonly wear no protective clothing, and it is not unusual to see them in their shirtsleeves on a warm spring day, hosing the roadside and engulfed in a cloud of herbicide. Perhaps protective clothing and strict precautionary measures would inspire misgivings about the chemicals in both the workers themselves and the public. Convinced that the herbicides are harmless, spray crews tend to be defensive, belligerent, and occasionally reckless toward concerned or alarmed citizens.

Houses in remote rural areas tend to be located as close to the road as possible, saving costs of access lines for phone and electricity.

"When you live out here alone and so few cars ever come your way – well, you just don't want to miss none of 'em," as one elderly farmer says.

Gardens, yards, and even front porches often adjoin the road, water supplies flow under it in culverts, open ditches, and pipes, and roads themselves often follow river banks.

For years, people had no basis for complaint when their property and water were sprayed by road crews other than indignation at the unpleasant smell and taste, the unsightly appearance of the brown, withered hedges and shrubs in their yards. Beekeepers who complained that the roadside spraying killed their bee colonies would be visited by county officials or agricultural extension agents, who would declare unequivocally that the bees had died of foulbrood or some other disease. Experienced beekeepers knew better, but had no resources to prove their claims. "NO SPRAY" signs put up by road-side dwellers were largely ignored.

Controversy over herbicides has centered chiefly on aerial spraying, while roadside spraying continues almost unchecked. Aerial spraying – especially in the uneven terrain and erratic climate of the Pacific Northwest – poses a serious threat to land far beyond the treated areas, caused by drift, vaporization, and the creation of toxic compounds (dioxins) in the burning of treated areas. Roadside spraying is performed by tank truck sprayers with less danger of aerial drift. But the herbicide mixtures applied are as much as twelve times more concentrated[9]. This intense concentration of chemical poisons is applied to miles of roadsides bordering private homes, school yards, farm land, pastures, and parks.

The black surfaces of roads and oiled shoulders radiate far more heat than either living or dead vegetation, vaporizing the poisons readily. Unwitting travelers from distant cities who may never have heard of herbicides drive for hours along state, county, and federal highways, enjoying a day in the country with no suspicion of what they might be breathing along the way. Often they are tempted by the blackberry jungles that thrive along cleared road-sides (sprayed relentlessly to no avail for years), picking buckets of them to take home for jam and preserves. State and county spray crews post no signs warning travelers of the danger, although some of the highest concentrations of dioxin (TCDD) in wildlife samples have been found in small birds and mammals collected along sprayed roadsides[10].

Children spend up to three hours a day in school buses traveling these sprayed roads every school day, breathing the vaporized herbicides. They wait for their buses at bus stops that are sprayed, and play on school playgrounds and sports fields that are often treated with the same chemicals. Truckers and others whose jobs require long hours on the road are also chronically exposed to these poisons.

The Durbins' water and sediment sample was one of only two taken by Dr. Savage in the Five Rivers valley. (Dr. Savage's second sample, taken several miles downriver, was negative for TCDD.) According to former millworkers, "people from EPA" also took samples from all water sources for the mill at Five Rivers. Although no results of testing were ever made public, workers say, shortly thereafter the mill owners ordered the drilling of two 300-foot-deep wells, one at the mill camp where the Durbins lived. Henceforth, the mill would draw its water supply from deep aquifers.

The TCDD component of 2,4,5-T is a long-lasting contaminant in the environment[11]. Whether the TCDD in the Durbins' water came from road-side or aerial spraying, or both, its presence is alarming.

"They simply can't tell us whether that's all there ever was in our water, or if that twenty parts per trillion is just the tail end of a much larger amount washed in there years ago," Irene says. (By some scientific estimates, chronic exposure to as little as .05 parts per trillion is sufficient to pose a grave threat to human health.)[12]In addition, the presence of TCDD is an indicator of how much 2,4,5-T and 2,4-D humans have been exposed to for many years. The water supplies of other families in Five Rivers have not been tested for TCDD, and may never be. Such analyses are prohibitively expensive. These families, their insurance companies, or public health programs will continue to pay a price of herbicide contamination never calculated in industry or Forest Service cost-benefit analyses.

"We moved from Five Rivers to Dallas, Oregon, in May of 1979," Irene says. "Since the move, the children have suffered no nosebleeds. The severe sinus attacks, headaches, and eczema have stopped. My husband's diarrhea has almost disappeared. He did have the heart attacks in the spring of 1980, but his family has a history of heart trouble, and we were all under considerable stress and anxiety at the time. He was only 36 years old then, though. That's a young age to have any kind of heart attack. There is just no way of knowing whether exposure to dioxin or the herbicides during all those years could have made him more susceptible in some way.

"Our son's respiratory problems have cleared up, but we still worry about him. He's slower than other kids his age. He seems unable to learn simple things like colors and tying shoes that they all can do at his age. He has what they call behavior problems – maybe because we overprotected him, thinking he wouldn't live. But I think there's more to it than that. I've known other kids who've been overprotected and fussed over – they may be bratty and difficult, but they don't have learning problems like this. If anything, they seem a little brighter than other kids."

A shower of water bursts through the screen window over the sink. The kids outside are playing fireman with the hose. Irene goes to the door and tells them the hose is to stay off for a while. If they will run around in the sun and dry off, she tells them, they can come in for graham crackers and milk.

"We both feel that the symptoms we had were more than just coincidence," she says. She pours milk into six paper cups and opens a box of graham crackers. "We are not by nature inclined to protest. But we care about our children, and about the other families who still live in the valley. Those are the best friends we ever had. We know what we've suffered, and what they have suffered. It's so hard to express that in front of outsiders, like at the EPA hearings, where those Dow lawyers try to force us to prove that our problems came from the spraying. We can't do that." (In 1980, Irene testified as a witness in the 2,4,5-T registration-cancellation hearings.)

Image by Rio Davidson
A narrow buffer of trees along a country road separates a stream from a clearcut

"Maybe we can't prove anything, but that doesn't mean that people aren't being hurt – and hurt seriously – by the herbicides. Their suffering is unnecessary. Somehow it should be stopped."

* * *

The Alsea valley, where Debbie Marano spent much of her childhood and returned to live in 1979, is a fertile, gently sloping pocket in the coast range, some twenty miles east of the conjunction of the Alsea River and Five Rivers. The town of Alsea is little more than a way station along two blocks of State Highway 34 as it follows the meandering Alsea River to the coast, forty miles away. There are a gas station, two small groceries, a body shop, and the weathered, antique homeliness of the Alsea Mercantile Co. The Farmer's Kitchen is the only restaurant for forty miles in either direction, and a cafe doubles as the local tavern. The Alsea School holds all grades in a single structure, its playgrounds bordered by unmown fields. There is no doctor, no dentist, and even the church is a mile from town. Crops and pasture begin abruptly behind the single row of businesses along the highway, as if grudging the shops even their narrow strip of frontage. It is an hour's drive to the nearest hospital, movie house, fast-food chain or department store.

South of Alsea, a gravel road runs eastward off the road through Lobster Valley. It is a logging road, primarily, that winds through uninhabited timber country of the Coast Range mountains, linking Alsea with the Willamette Valley. No power poles nor telephone lines border this road. The few scattered private holdings in the area have neither electricity nor phones.

Traffic along this road during the day consists mainly of log trucks and "Crummies" carrying forest workers to their jobs, with a few local vehicles and occasional travelers who prefer back roads. At night, however, any traffic at all is rare, and the car seems an intruder from a future time. There are no lights anywhere, no sign except the road itself that man has ever set foot here. On this cold night in late winter, as you pull off the road and shut off the motor and lights, there is nothingness, a blackness that is frightening.

Few driveways turn off this road, and there are no mailboxes. The area is too remote and the population too sparse for mail delivery. Almost obscured by brush, a small, handpainted sign on one of the rare driveways says, "LAMP

LIGHT TROUT FLIES." At the bottom of the steep, rutted drive, dark forest opens into meadowland and orchard. The windows of the single house glow with the light of kerosene lamps. Bonnie and Tony Hill are at home.

Bonnie is a schoolteacher in Alsea who wrote a letter in 1978, along with Debbie Marano and six other women from the Alsea and Five Rivers areas that had been heavily sprayed with herbicides for many years. The women wrote of the miscarriages they had experienced following spray operations near their homes. They sent a copy of their letter to EPA, which initiated a study of "spontaneous" abortions in the area. As a result of that study, EPA issued an emergency suspension order on 2,4,5-T and silvex for most uses, including forestry.

The Hills' house is warm and bright after the cold darkness outside. Around a lamp at the kitchen table, their three children are doing their homework. Near them another lamp lights Tony's fly-tying table, with its odd assortment of tools, a special vice, tufts of animal fur, deer tails, iridescent neck hackles from different kinds of roosters and other fowl. On the walls, glass cases display finished flies like jewels or rare butterflies.

Bonnie is a soft-spoken woman in her late thirties, slender, with graying hair pulled loosely back from her face. "The CATS case was no more than a vague background to my involvement with herbicides," she says. "I really didn't know very much about it, just the little that I read in the papers, which never discussed toxicology.

"The climate in Alsea at that time was such that the CATS people were seen as just a bunch of radicals and freaks who were living out there in Five Rivers. I had students in my classes whose fathers were employed by timber companies or the Forest Service, and I remember the comments they would make about the 'crazies' out there.

"Then in the summer of '77 I came across a packet of information published by CATS at a crafts fair. I bought it, because it included studies on the toxicology of herbicides. One thing I had learned from the CATS case was that herbicides were used in our area. I took it home and read it, and came across a synopsis of Allen's studies with monkeys in Wisconsin[13]. That's what really started my wheels turning – when I read about spontaneous abortions in those monkeys.

"Of course the first question to come to my mind was whether or not these chemicals had touched my own life. I had had a fetal death – a spontaneous abortion – in 1975. As I thought more and more about it, I remembered other women, former students of mine. They were young and had married after graduating, and still lived in the area. I still saw them from time to time, and I remembered them talking about miscarriages they had had. As I thought about it, I could think of three who had miscarried in the springtime. It seemed really odd that there were so many of us who had miscarried exclusively in the springtime.

"I spoke with each of these women about their miscarriages.

"They in turn told me of other women they knew, and I went to see them. That whole process took me a long time, because I was teaching full time and I had my family to care for. By the spring of '78 I found eight women who had eleven miscarriages altogether. They all occurred in the springtime. I hadn't found any that occurred at any other time of the year. And I asked – I wasn't selectively choosing incidents that I knew about. Every single one that I came across happened in the springtime. That seemed pretty unusual.

"At that point, I decided to gather what information I could about sprayings in the area. I made phone calls to the Bureau of Land Management and went to the Forest Service and to Starker and Willamette, the timber companies that own land around here. I tried to learn when they sprayed, where, and what compounds and amounts they used.

"I got varying degrees of cooperation. Some people were very reluctant to talk. A whole section of my testimony to EPA documented the reasons I think the spray data even now is incredibly incomplete. When I saw the spray data EPA had put together – the total poundage of herbicide use they calculated for this area – it was just incredible. In some cases I had figures from these various companies that exceeded EPA's figures, and I know I didn't have complete figures. It was just amazing to me that a federal agency hadn't even come close.

"I gathered spray data as best I could. I mapped where the women lived in relation to the spray sites and put it together on a chart. It certainly appeared to me that there was a correlation between the timing of the spraying and the miscarriages that occurred nearby, but only a much larger study would show if that was a consistent pattern. I decided at that point that obviously our resources were too limited.

"I talked to the other women about getting someone else in to look at the situation to see what they could come up with. We wrote a letter telling what we had found, made a bunch of copies of it and the chart, and sent it to everyone we could imagine that might be interested. From the beginning it was our idea that whoever looked at it would look at more than just us eight women. Hopefully they would look at the whole area to see if this was a general problem, not just limit it to the eight of us.

"When EPA responded, I wasn't aware that people all over the country had been making the same kinds of observations for years and years and years, and that there was this long history of complaints about phenoxy herbicides. What made them respond to us, to the women in Alsea? I don't know. The CATS case paved the way somewhat, I guess, and Vietnam. I think the media might have had an influence, too, in the decision to look into the Alsea situation, and that was a fluke. We didn't seek out the media. That wasn't part of our plan, to have national news coverage. It just happened. Overnight, there seemed to be all this interest,"

* * *

What Bonnie did not know was how very close the Alsea women's letter came to going the way of all letters to federal agencies. In February of 1978, two months before the letter was sent, EPA had declared a Rebuttable Presumption Against Registration (R.P.A.R.) of 2,4,5-T, meaning that unless the manufacturers could come up with new data to establish the chemical's safety, its registration would not be continued. Within a month of the R.P.A.R. announcement, EPA had received over 4,000 letters from the public, commenting on its action. Bonnie and the other seven women were unaware that their letter had been tossed into a pile of these responses, where it would undoubtedly have passed unnoticed.

However, their letter caught the attention of a reporter, who suggested to Erik Jansson, a research associate for Friends of the Earth in Washington, D.C., that it might be of interest to him. Jansson went through all the R.P.A.R. response letters, from which he collected over 450 poisoning incidents involving 2,4,5-T.

When Jansson came upon the Alsea letter, he was immediately impressed. It was carefully documented, well summarized. The women made no bold-faced claims, but merely stated what they had been able to find out and asked for help in finding out more. Their letter was brief – one page, plus a chart. Jansson knew how little chance the letter had of being noticed in the heaps of R.P.A.R. responses. He had many copies made of the letter and "papered D.C. with it," as a friend of his commented. He made certain that it got into the hands of media people and legislators, demanding that the women deserved a full-fledged investigation.[14]

EPA was not responsive at first, but eventually the public interest Jansson stirred up embarrassed the agency into looking at the problem. EPA sent a team headed by Dr. Eldon Savage of the Colorado State University's Epidemiological Pesticides Study Center to conduct a study. The "study" consisted of interviews with the eight women who had signed the letter. No attempt was made to examine miscarriage rates in the general population of the area.

The Benton County health officer, Dr. Norton Kalishman, as well as groups concerned with the herbicide issue, such as FOE and CATH, objected strenuously to a "study" based solely on oral interviews with a limited number of subjects. "There were a group of civil service employees at EPA at the time who wanted to leave 2,4,5-T on the market while they 'monitored' exposed populations for more poisoning episodes as they turned up," Erik Jansson said. "Along with some of the EPA Carter appointees, I hit the ceiling, saying that this amounted to experimentation on human beings."[15]

"Those first questionnaires were inadequate," Dr. Kalishman said. "Alsea II – the second study – was done because we yelled and screamed. I was awful. I was rude. I ranted at them over the phone, bullied them into it. That's the only way you can get them to move ..."[16]

Pressured by Dr. Kalishman, Erik Jansson, environmental groups, and its own personnel, EPA instituted what is now known as the "Alsea II study," an attempt to study spontaneous abortion rates in the general population of the Alsea area in relationship to herbicide use. The study compiled spray data from BLM, the Forest Service, and private timber companies, and collected information on spontaneous abortions from hospitals and private doctors who agreed to participate (many refused), locating patients by their zip codes. It pooled data from sprayed areas with data from urban centers. Two control areas, presumably not exposed to 2,4,5-T, were selected for comparison: an urban control area in the central Willamette Valley of Oregon, and a rural control area in Malheur County in eastern Oregon, where little 2,4,5-T use was reported (although 2,4-D was widely used).

When data from all areas for all years of the study were pooled, there was a significant peak in the month of June in the study area that did not coincide with the control areas. The June peak in spontaneous abortion rates represented a two-month shift from the approximate time of application of forest herbicides, which is generally consistent with animal research as far as the timing of herbicidal effects on fetal survival. The study area had a spontaneous abortion rate at its peak of 130 per 1,000, whereas the control area had 46 per 1,000.

It was on the basis of this study that the EPA issued its emergency suspension order against 2,4,5-T and 2,4,5-TP (silvex) on February 28, 1979, effective immediately for all uses except rangelands and rice crops. At a press conference on March 1, 1979, EPA Deputy Administrator Barbara Blum announced:

We have just received the results of a study that shows a high probability that the herbicide is linked to actual human miscarriages in an area where 2,4,5-T and silvex are used regularly ...

New studies in the Alsea basin of Oregon show a high miscarriage rate shortly after the spraying of 2,4,5-T in the forests. This alarming correlation comes at a time when 7 million pounds of 2,4,5-T are about to be used across the nation to control weeds on powerline rights of way, to manage forest lands, and to control weeds in pastures. The emergency suspension action we are taking today will protect the nearly 4 million people who may be unknowingly and involuntarily exposed as a result of those uses.[17]

* * *

Inexplicably, the suspension did not apply to rangeland and rice crops—major routes of exposure to the herbicides and their dioxin contaminants for the entire nation – in spite of the fact that dioxin had been found several years before in beef fat and human milk from rangeland areas of Texas where 2,4,5-T was used[18].

The emergency ban on 2,4,5-T and silvex was to be effective until suspension hearings in April determined whether the EPA would proceed with lengthy cancellation hearings, which could take up to two years. At these hearings, manufacturers of the chemicals would have to present convincing evidence of their safety and benefits to be weighed against evidence – both scientific and "anecdotal" – of their potential harm.

Dow Chemical immediately sued in federal court for an injunction against the ban. The injunction was denied on the grounds that it was not the court's jurisdiction to rule on the performance of a federal agency in fulfilling its mandate. Dow's next move was to withdraw from the suspension hearings three days after they had begun, along with all but one of the other plaintiffs, essentially leaving EPA in limbo. What Dow and other manufacturers wanted was to proceed directly to cancellation hearings, which would shorten the time the emergency ban was in effect if the manufacturers won their case.

The Alsea II study, which justified the emergency suspension, was attacked vehemently by both industry and environmental groups. Both sides raised essentially the same criticisms:

- Data for the "study area" pooled information on spontaneous abortions from both sprayed and unsprayed portions of the area. No attempt was made to locate possible "Clumping" of miscarriages in particular locations (such as the heavily sprayed Alsea-Five Rivers watershed).
- The study did not determine what other chemicals the populations of the study area had been exposed to, in particular 2,4-D, which was commonly applied in combination with 2,4,5-T and silvex.
- The spray-use data (locations, amounts, times, concentrations, and acreages) collected for the study were inadequate.

- The study focused solely on spontaneous abortions, which are only one aspect of a whole range of interrelated reproductive functions known to be affected by chemical insult, depending on timing, individual susceptibility, fetal development, and other factors. The study should have considered such related problems as uterine hemorrhaging, hysterectomies, inability to conceive, tubal pregnancies, stillbirths, birth defects (particularly neural tube defects), all of which had been reported in the sprayed areas.
- The study included no data from private doctors or clinics.
- "Most spontaneous abortions occur in the first trimester, and many women don't even go to a doctor," Bonnie Hill says. "Only two of the women I originally surveyed went to a hospital. So the hospital records were just the tip of the iceberg."[19]

* * *

These substantial criticisms of the Alsea II study were clouded by an official critique provided to EPA by Oregon State University, assaulting the study for lack of scientific rigor. In fact, in strict scientific terms, no study of human effects from field use of toxic chemicals can ever be conclusive.

As Dr. Theodore Sterling points out:

> If your animal studies are inadequate or questionable and you have a chemical that is affecting humans, you have a situation where you are forced to study its effects on people, and there is a lot of difficulty in doing this. The Alsea study is a good example. The Alsea study is as good a study as you can do, of this type. You can jump all over any study that's not done under laboratory conditions, because we cannot do a perfect study of human beings. There are just some studies which are impossible to do clean.
>
> Look at the effects of radiation on offspring, for example. You cannot just take a random sample of pregnant women and irradiate them. You just don't do that. So you rely on finding a group like that, which is very difficult, and when

you do find such a group of pregnant women who have been irradiated it will not be a random selection, it will already be limited to a particular situation and circumstance. All you can really do is work your way around the data to come to some reasonable conclusion whether there is an effect or not, and whether or not it's the radiation that causes the effect, but there's no way in the world to design such a study and implement it when you're dealing with human beings.

It's the same way with the Alsea study. You have no way of designing a study because spraying doesn't come in an exact area, you don't even know what the dose is, you don't even know whether your control group doesn't also have some exposure. It's a mess. But there are no health studies done on this, so you just have to sit down and try to draw conclusions from what you've got. If you're not willing to draw conclusions from any study, then why do any? In that framework, the Alsea study is a decent study.[20]

"The Alsea II study is really only important with reference to the emergency suspension," a CATH spokesman noted. "In order to issue a suspension order, the EPA administrator must have strong evidence that adverse human effects are occurring from a particular chemical. The political decision was made to take 2,4,5-T off the market, and then EPA came up with a fifty-cent study to justify the decision."

Dow and other manufacturers prevailed upon EPA to proceed directly to cancellation hearings, by-passing lengthy suspension hearings. With the ban in effect, cancellation hearings on 2,4,5-T and silvex finally began in early 1980. The hearings were to be divided into two parts, a "risk" section to assess testimony concerning the hazards of the chemicals, and a "benefits" section to hear evidence of their efficacy and economic necessity. An administrative law judge would weigh the risks against the benefits and submit a recommendation to the EPA administrator. Bill Bitsas, network coordinator for Northwest Coalition for Alternatives to Pesticides (NCAP), put it another way. "What it means is, they'll weigh the benefits to [Dow] executives in Midland, Michigan, against the risks to the people of Oregon."

The risk section opened first and took nearly a year to complete. Testifying in support of cancellation were Bonnie Hill, Salty and Gisele Green, Dr. Elam, Irene Durbin, and others from the Pacific Northwest as well as physicians and scientists from Oregon and other parts of the U.S. The benefits section was to have begun in early 1981.

With regard to forestry, industry would have to establish that the economic and social benefits of herbicides outweighed any risk the chemicals imposed. Herbicides had long been justified as a forest management tool on the assumption that they increased timber (conifer) growth and yield by eliminating competing vegetation and were far cheaper and more effective than alternative methods of weed control. Based on that assumption, the Forest Service and Oregon State Department of Forestry predicted that 20,000 jobs would be lost in Oregon alone if 2,4,5-T were banned. The overwhelming "benefit" of herbicides was that they were essential to the timber industry, the economic base of the Pacific Northwest.

EPA intended to present strong evidence in this section that, indeed, few if any actual benefits accrued from herbicide use in forest management. EPA's key witness in this section was Dr. Jan Newton, an economist from Eugene who worked as research consultant and contract researcher for governmental agencies, such as USDA and the Department of Health, Education and Welfare. One of her recent projects had been a year-long study of herbicide use in forest management, evaluating the scientific adequacy of the economic base for herbicide use. Her study entailed review of over 8,000 pages of research findings, public documents, and articles.

What Dr. Newton found was that no research had ever confirmed that herbicides enhance growth in timber trees. The studies often cited to indicate herbicide effectiveness were inadequately designed, most often did not involve aerial applications (the studies were done on backpack ground spraying of individual plants), and did not assess or take into account the damage herbicides inflicted on conifers. She found that the cost figures given for herbicide use were entirely erroneous. She concluded that future timber-yield projections based on nonexistent and faulty research were therefore meaningless. The only "benefit" of herbicide use in forestry she could identify was that their fictional enhancement of timber yield justified increases in the harvest of timber currently allowed on public lands.[21]

In terms of the "risk/benefit" analysis which would determine the outcome of the cancellation hearings, Dr. Newton's testimony would be particularly damaging to industry's case. The lack of any firm evidence of benefit would make any risk outweigh alleged benefits. Dr. Newton's research would establish on the record that neither 2,4,5-T nor silvex had any proven benefits in forestry, casting doubt on the efficacy of all herbicides used in forest management. Perhaps for this reason, the benefit portion of the hearings was repeatedly postponed.

In late spring, 1981, soon after the Reagan administration came into office, Dow Chemical Company began secret negotiations with EPA for a settlement of the 2,4,5-T issue that would preclude continuing the hearings.[22] Those negotiations were conducted entirely in secret, despite vigorous protests by NCAP and other environmental groups. The benefits section of the cancellation proceedings has yet to be heard. If the history of Dow and EPA with regard to 2,4,5-T is any indication, the suspension of 2,4,5-T will be lifted and its registration continued without any evidence of its benefits – or lack of them – ever being presented. At hearings before the Subcommittee on Forestry of the U.S. House Agriculture Committee in April 1981, Lane County (Oregon) Commissioner Jerry Rust told Congressman Weaver that to bring 2,4,5-T back on the market "would be a declaration of open war against the people of Oregon."[23]

* * *

The house is still and quiet now. Only the lamp over Bonnie's papers on the living room table and the lamp by Tony's workbench are still lit. The children have finished their homework and gone to bed.

"It's been a puzzle to me all along," Bonnie says. "Why would a billion-dollar industry fight so hard and spend so much to keep a few chemicals on the market when so many people have apparently been harmed? Their credibility will be affected. People will be nervous about other kinds of chemicals. They will see the discrepancy between their own experience and industry's claims. The only explanation seems to be industry's belief in the domino theory: if one chemical like 2,4,5-T goes down a whole lot of others will follow[24].

"If that's true, though, it just means they were all standing on the same shaky ground, doesn't it? Maybe Dow is right when they say 2,4,5-T is one of the safest products they make.

"On a local level, the resistance is more understandable. There are all these applicators and users who have been using these substances on trust, in the belief that what the agencies and manufacturers said about them was true. They thought they were safe and did the best job. When you try to talk to them about problems with the chemicals, I think they really feel threatened. They react so adamantly, so intensely, because they can't admit to themselves they may have used something that has caused real harm to people. Not that anyone's trying to blame them for it, but they would blame themselves. I think they overreact because they can't begin to admit they may have made a mistake – because of the implications. What did that mistake mean? It's too mind-boggling for them."

Bonnie spreads newspaper clippings about the suspension order on the table, along with a copy of the Alsea II study and the OSU critique of it.

"When industry sees a rising tide of public concern and sentiment over the use of toxic chemicals in a particular area, it becomes really important to them to squelch it, to give people enough misinformation that they will be confused about the real issue. The tactics are so bald – identifying anti-herbicide people as dope-growers and welfare hippies, publicizing spurious statistics showing how many thousands of jobs will be lost if these chemicals are banned, presenting misleading data from so-called scientific sources like OSU.

"That's one thing that really shocked me when I became involved in all this, seeing the role of OSU in the herbicide question. I wish somebody could do an investigation of OSU, track down their grants. But people who try keep running into something called the OSU Foundation. Companies, like agricultural chemical companies, can give money to the Foundation and the foundation promises-somehow it's legal to keep the donors anonymous if they request it.

"It's that sort of set up, I think, that allows things like this to happen, and what can you do? Look at Michael Newton. He can make statements to the press like the one he made about no herbicide residues ever being found in mammals exposed to herbicides in the coast range. Even though the EPA had already found residues!

"And look at the OSU critique by OSU scientists of the E.P.A's Alsea II study. They have a whole song and dance about the study area that EPA looked at – a sixteen hundred-square mile area. OSU says this is not 'forested rural area' but is really 'urban coastal area'. They call us an urban population. They say they got their figures on the population out of the Oregon Blue Book, the one the Oregon Secretary of State brings out. But they—OSU— were just totally wrong. When you look at the Blue Book you can see a certain set of figures, and when you look at the OSU report the same set of figures is completely different. They're just incorrect."

Bonnie leafs incredulously through the OSU critique.

"Another thing they did was to determine that a person was an urban resident if they had a post office box. According to OSU, if you have a post office box, you're an urban resident. Well, we have a post office box – because we live so far out that we can't get the mail delivered. So we are supposed to be 'urban' residents, even though we don't have any neighbors for twenty miles on one side of us and a mile and a half on the other. It's incredible. How could they say that? But they are the scientists, they are the ones EPA will rely on.

"The other thing about the whole issue that just astonished me – I know I was naïve – was finding out about trade secrets. That's something that people just don't want to believe-that this kind of thing goes on. Industry can withhold information about a chemical's ill effects on lab animals, and can classify that information as 'trade secret'. This means the public has no way of ever knowing the effects of a chemical. In spite of a Supreme Court decision to the contrary, EPA will not release the information. I just can't believe that. I just can't believe that could go on in this country."

She opens several issues of NCAP News to articles on the subject.

"Look at Roundup, Monsanto's answer to 2,4-D. You hear and see it advertised everywhere. But the people who are living here, being exposed to Roundup, have not been able to find out anything about it because all the tests are classified as trade secrets."

(The U.S. Forest Service had sprayed Roundup in the Siuslaw Forest since 1978 on an experimental-use permit, as the chemical was not registered for forest use at that time.)

"Then we learned that former officers of the laboratory that tested Roundup are under indictment by a federal grand jury for deliberately falsifying test data. When we followed up on that, we learned that one of the men indicted has headed Monsanto's toxicology research since 1972[25]. Yet they are allowed to go on spraying Roundup. There just seems to be something basically wrong.

"Industry makes it as hard and expensive as possible for anybody to fight these things. They want to show us that this is going to be an incredible battle. Any time we raise questions about one substance, it's going to be years and years and years of people devoting all their time and energy and money to fight a product on the market, and there'll be twenty more registered in the meantime, just as bad or worse.

"They've shown us how hard they can make it for us. They've gone to great lengths on things like the Lincoln County herbicide measures to control public opinion. They are worried about public opinion. I think they're genuinely afraid of people finding out what's really going on. If they weren't afraid, they wouldn't spend so much to put out misinformation and to conceal the real effects of chemicals. If they had nothing to hide, they wouldn't fight so hard to keep the real information secret. But the bigger an issue they make of it, the more people are at least going to suspect that there is a problem."

There is a sudden whimpering from one of the darkened bedrooms. Tony gets up and carries a sleepy baby back into the lamplight for reassurance. He helps her drink a glass of milk. She falls asleep again, her milky chin nestled in a sleeve of his wool shirt.

"It's hard when you already have a full life to get involved with something as big as this. We've had hard times at home, to be sure. It hasn't been a problem between Tony and me, because he's been so supportive and so involved himself. But for the kids, the long-term effect of my devoting so much time and energy to the herbicide issue was that it came out of time I ordinarily spent with them. That's been hard. The most concentrated, intense effort I spent was preparing my EPA testimony, which ended up being 45 pages long with about 70 pages of supporting documents. It really took me a long time. It took a lot of research and was really an intense effort. At one point I was driving Jason-our eleven-year-old-to school for something, and he said, 'Mom, is it too late to get out of this?'

"Yet there are some really valuable lessons for them, too. I don't regret it happening at all, because it's good for kids to see their families involved in something that has repercussions beyond their own world, beyond their own immediate lives. I think Jason isn't really regretting that it happened. At that time he was just feeling all the pressure I was under. I think it's good for them to see that there are certain things we will stand up for, and certain things we will not abide. That's a real important lesson to learn.

"There are people in Alsea who are really glad the issue has finally been raised, I think. I can sense that by the nonverbal contact I have with them. When I go to the post office, matrons who are fifty years old and raised their families in Alsea look at me, respond and talk to me in a way that says, 'I know what you're doing and I really appreciate it. I think it's important and I'm glad you're doing it because I never could.' But they would never say that. Their husbands work in the woods, some of their husbands own the logging companies, and they just cannot speak of such things themselves. But they're glad other people are.

"It's a big problem to be publicly involved in anything political. There's even a bigger problem when the media shows interest. That was certainly a problem in Alsea. None of us felt comfortable being the center of attention. The first time a TV crew came to our place, there were two of the other women here. We had been just petrified for days. Whenever anybody from the media came, we spent several days of intense anxiety. Yet the strange thing is that once they were here, they were just people asking questions. I felt so full of a desire to let people know what was happening – you know, people needed to know about this – that once I got started, once people were here asking questions and showing an interest in the issue, it wasn't so difficult.

"But the hardest thing about talking with the media people was knowing how sensitive our own community is to the issue. It's real scary, not knowing how your own community will react, and knowing how they've reacted to other situations in the past. It's a big unknown whether you will be ostracized or whether people will still be able to accept you for the person that they've known for a long time.

"We've lived here ten years now. My father was a doctor in the Air Force, so we traveled a lot when I was growing up. Having a home, a community I belong in, is important to me. I'm very concerned about the way people think of me, but I couldn't see something so very wrong and let that keep me from doing something about it. I've invested a lot of time and energy in this community. These people are very important to me. That's why we're staying here. We like the people. But the issue was important enough to risk some of that."

Tony carries their daughter off to bed. He and Bonnie walk us to the car. Outside, the grass crunches under our feet, the windshield is white with frost.

"We came here because we'd had it with urban life," Bonnie says. "We were attracted to the Pacific Northwest because of the geography – the mountains and trees and rivers. We chose Alsea because of the people here. If we've accomplished anything, it's for this place and these people as much as for ourselves. My education made such a difference, just in knowing about resources. When something happened, we had some idea how to find out about it. That's the important thing – finding out and letting other people know."

Poisons: Innocent Until Proved Guilty

Give me the liberty to know, to utter, and to argue freely
according to conscience, above all liberties.

—John Milton

In 1972, ten years after the publication of Silent Spring, Congress added new provisions to the pesticide law requiring that pesticides must be tested for efficacy and for adverse health effects before they can be marketed in the United States.[1] Similar testing is also required for drugs. Manufacturers are responsible for such testing, which provides the basis for regulatory decisions by the Food and Drug Administration on drugs and by EPA on pesticides. The regulatory process involves weighing the risks against the possible benefits of use of potential poisons.

In theory, the democratic process allows the public to assess risks and decide whether or not to accept them. However, if the public is denied access to accurate knowledge – because government and industry withhold information and conduct research in secret – then its consent is meaningless.

Public alarm over herbicide safety has arisen since the late 1960s primarily from independent scientists and from growing ranks of victims, whose experiences shattered decades of public trust in industry and government experts. Veterans, parents, farmers, and other affected citizens have battled in court, in legislative hearings, administrative appeals, petitions, public meetings, and every democratic mechanism available, not only for their rights to be recompensed for damages and to avoid exposure, but also for their right to be told the truth about the poisons affecting them.

Ideally, pesticides should be adequately tested for safety and long-term effects before they can be registered and used. As has been documented ear-lier in this book, it has become increasingly clear that for many years poisons such as herbicides have been registered, manufactured, and used without suf-ficient knowledge about their effects on humans or the environment. Many pesticides were registered prior to the establishment of strict testing require-ments[2]. The Federal Insecticide, Fungicide and Rodenticide Act, which gov-erns the regulation of pesticides, places the burden of proving safety on the manufacturer (registrant) of a chemical, but also says that once a pesticide is registered, EPA must have clear and convincing evidence of harm to humans or the environment before moving to restrict or cancel its registration. FI-FRA specifically provides that the mere fact that a pesticide is registered is to be construed as prima facie evidence that it is safe when used as directed. In other words, registration is the magic formula that shifts the burdens of proof from manufacturer to EPA and the public.

If registration were indeed based on thorough testing and evaluation of all available knowledge, and if registrations were regularly reviewed as more information became available, the process might be an adequate means of assuring the safety of pesticide poisons. However, amendments to FIFRA in 1978 added an interesting loophole called conditional registration, which allows manufacturers to register their products without completing health and safety studies first. In addition, manufacturers are allowed to submit mere summaries of experiments rather than entire studies for evaluation by EPA. Like a blind man trusting the horse trader, EPA bases its regulatory decisions not on actual test data, but on the manufacturers' summaries and conclusions. Improper, nonexistent, or fraudulent safety studies may support registration of a chemical, and FIFRA does not provide for immediate and arbitrary suspension of a registration in the event that fraud or inadequacy is later detected[3].

A manufacturer may thus submit fraudulent studies to EPA with confi-dence that its product will be securely registered before the fraud is detected, especially since the studies themselves are protected by trade-secret policies and immune from public scrutiny. Once the product is registered – even after fraud is detected – the burden of proving it unsafe rests with the public. The registered chemical is innocent until proven guilty, and the public has scant recourse in law.

The same companies often manufacture both pesticides and drugs, and the same independent testing laboratories are commonly employed to test them. Adverse effects of drugs on human health are more readily apparent than effects of pesticides, because drug exposure is known and often monitored by physicians, whose observations might be at variance with fraudulent studies. The discovery of widespread fraud and inadequacies in drug testing is an indication of similar, and perhaps greater, problems in the testing of pesticides.

In the case of thalidomide, for example, the research on reproductive effects of the drug had simply never been done, although the company's promotional material to doctors was cleverly worded to imply that it had. When the FDA requested further information from Richardson-Merrell Drug Company (the North American distributors of thalidomide), the company submitted a fraudulent study which had been published in the American Journal of Obstetrics and Gynecology. The study was written not by the physician who lent his name as author, but by the medical director of Richardson-Merrell and his secretary. The four "thalidomide babies" delivered by the physician "author" were not mentioned in the study, and all data were reported to have been destroyed.[4]

The destruction of experimental records which would prove or disprove manufacturers' conclusions of safety tests is a common occurrence. "It is amazing how many times what we refer to as the Andrea Doria phenomena occurs," says Dr. Alan B. Lisook of FDA, citing some of the reasons given to explain why records were not available, as follows:

- They were destroyed in a fire.
- They were destroyed in a flood.
- They were dropped in a sewer and had to be destroyed because of the stench.
- They were lost in a boating accident.
- Burglary, robbery, or vandalism.
- The hospital closed and the records were lost.

- They were lost in the mail.[5]

Both Richardson-Merrell and Grünenthal (the German company which developed thalidomide) persistently denied any connection between the drug and the birth defects and crippling neurological problems associated with it, in spite of massive epidemiological evidence. The stubborn skepticism of a single FDA pharmacologist was all that prevented thalidomide from being approved for over-the-counter sales in the United States.

In the early 1960s, officials of Richardson-Merrell were indicted by a U.S. grand jury for submitting false and fraudulent information to FDA about another drug, Mer-29, and withholding information about its devastating effects on laboratory animals. After FDA had approved the drug, the company consistently denied "anecdotal" reports of severe side effects from physicians, patients, and even other drug companies during the time the drug was marketed. Mer-29 was an anticholesterol drug that caused cataracts, baldness, skin irritations, nausea, and vomiting. It was marketed with FDA approval for 22 months before a chance meeting between an FDA inspector and a telephone-company employee, whose wife had worked in Richardson-Merrell's lab, led to an FDA "raid" on the company's lab in Cincinnati in 1962. Two days later, Richardson-Merrell withdrew Mer-29 from the market[6].

The association of often-fatal Toxic Shock Syndrome with certain tampons provides another example of an industry campaign to convince the public that its products are safe, against mounting "anecdotal" evidence that they are not. Few women were willing to wait for conclusive scientific proof that Toxic Shock Syndrome was not caused by the tampons; acceptable alternatives existed, and sales of the suspect tampons plummeted[7]. Faced with a choice between risk to health and slight inconvenience, consumers chose the inconvenience.

Inadequate or false safety information on drugs is likely – as casualties mount – to be refuted by physicians' and patients' "anecdotal" reports. Consumers then have the choice of whom to believe – the salesman or the victim – and the subsequent choice between using or not using the product.

Shoddy or falsified tests on pesticides are not so readily detected, and potential victims have little choice on the matter of exposure. Pesticides contaminate not only air and water, but foodstuffs, wood, clothing, and other products which are transported far from the origin of exposure. Affected populations are widely distributed, and are exposed to such a stew of varying chemicals (in addition to drugs and industrial wastes) that no single substance can be linked with certainty to a particular epidemiological effect. (As evidence of thalidomide's side effects began accumulating, one Grünenthal representative had even proposed mixing it with other drugs so that "if it proves impossible to keep things dark or to ward off attacks, any alleged side effects could then be attributed to the other preparations."[8]

Fraud in drug testing is shocking both because its disastrous consequences are identifiable and because it occurs in an industry supposedly dedicated to health and quality. The pesticide industry – associated with sophisticated methods of killing and historically linked to chemical and biological warfare research – does not enjoy so noble a reputation. In theory (and generally in practice) pharmaceuticals have a built-in monitoring system in the medical profession which prescribes them and can observe their effects. Pesticides have no such watchdog, and most physicians are woefully ill-educated in the complex field of chemical toxicology. Unless a chemical's effects are drastic and immediate in an identifiably exposed population, its long-term, sub-lethal effects may remain undetected – genetic damage that may not manifest itself for several generations, for example, or cancers which will not mature for 20 or 30 years.

In 1976, quite by accident, the FDA discovered discrepancies in testing performed by the nation's largest test laboratory, Industrial Bio-Test Laboratories (IBT). IBT, a subsidiary of Nalco Chemical Co., had conducted over 25,000 tests on many different products, ranging from food additives and colorings to pharmaceutical products, cosmetics, and agricultural chemicals, including forest herbicides. On the basis of IBT data, such products were registered for use by EPA and FDA, and residue tolerances of pesticides on foodstuffs were established.[9]

FDA's audit disclosed deliberate falsification of testing reports in many of 4,300 tests involving 123 pesticides and 160 applications for residue tolerance levels (the amount of a chemical allowed to be present on foods – fruits, vegetables, meat – for the market). An investigation of IBT was begun by the FDA, the EPA, the National Cancer Institute, and the Department of Health, Education and Welfare (HEW).

The investigation revealed widespread problems. Healthy test animals had been substituted for animals that became ill during the tests. Rats which had died of neglect or exposure to experimental substances were listed as having been killed for autopsy, and autopsy reports did not record the real cause of death. Records were incomplete. Tumors listed in the raw data on some studies were not recorded in the final reports. ("Raw data" include all the information recorded during the progress of an experiment: daily records of dose levels, weight measurements, examinations, births, deaths, pathology reports, chemical analyses. On the basis of this information, a final report is written, describing the effects noted and drawing conclusions about toxicity. For purposes of registration, EPA looked only at summaries of these final reports and therefore had no means of checking their accuracy or validity.)

In other IBT tests, rats were recorded as having died twice. Vital statistics continued to be recorded on test animals long after they had died. A former IBT technician called one study "the magic-pencil study," because complete blood analyses appeared in the final report, although most of the analyses had never been performed.

"The uproar grew louder," The Wall Street Journal reported in 1978, "when IBT said that [former IBT president] Mr. Frisque had inadvertently ordered the shredding of hundreds of records requested by EPA. Among the shredded data were seven long-term studies on the cancer-causing potential of substances used in plastics, herbicides, and cyclamates, the artificial sweeteners."

A 1977 EPA press release inexplicably labeled these serious shortcomings "deficiencies," as if the problem were a minor one that could be easily remedied. "The Environmental Protection Agency has found 'deficiencies' in some of the pesticide safety tests of a major laboratory that has done several thousand studies for pesticide manufacturers on whether their products cause ill effects, such as cancer, birth defects, nerve damage, or metabolic problems," the press release stated.

Further on in the press release, however, a statement by EPA Administrator Douglas M. Costle revealed the depth of EPA concern over the IBT "deficiencies."

"The discovery of these deficiencies in pesticide safety tests is a serious matter," Mr. Costle said. "EPA is dependent on receiving accurate test data in its registration of pesticides. If that data is deficient, there can be no assurance that the public health and the environment is being properly protected."[10]

As a result of the IBT investigation, EPA asked 235 chemical companies to re-examine 4,363 tests supporting registrations of 483 different pesticides. The pesticides in question included all herbicides currently used on the Siuslaw National Forest[11]. While tests are evaluated and retesting performed, registrations on all the pesticides tested by IBT continue unaffected.

"There is no excuse for EPA's reliance on fraudulent data to establish residue tolerances, and there is no excuse for its negligence in taking no action when the falsification of such data was brought to its attention," said Dr. Richard L. Doutt, a professor emeritus of entomology at the University of California. "Although the scientific integrity of tests on which EPA-based decisions on pesticide safety had been undermined, EPA took no action to amend in any way the registration of the affected pesticides. Furthermore, although EPA knew the results of the audit in 1976, it withheld information from the public until August 1977, after the congressional summer recess had begun, thereby avoiding critical review by hostile legislators.

"By March 1978, EPA officials confirmed that the test results had been deliberately distorted but no action was taken; no one was prosecuted. The credibility of EPA has been severely damaged by this episode and its inaction is inexplicable."[12]

News accounts of the IBT scandal published in 1978 gave the public little indication of the scope and implications of the fraud. After a small flurry of news items in 1977 and 1978, the subject of IBT faded from public view, while EPA reviewed thousands of test reports and bargained with manufacturers to provide new data. Along with twenty-two other foreign countries, Canada relied heavily on U. S. testing programs to establish its own guidelines for pesticide marketing. Canada and the U. S. therefore agreed to share the task of reviewing the IBT data.

Pesticide manufacturers were required to audit the raw data and conclusions of IBT studies they had submitted in support of product registrations and tolerances, and submit their validations to FDA, EPA, and Canada, which would then audit the industry validations, and determine which studies needed to be replaced to fill "data gaps." During this time, EPA, the pesticide industry, and industry lobbying groups, such as the National Forest Products Association, supported amendments to the pesticide law which were passed by Congress in 1978, specifically addressing the problems IBT posed to both industry and EPA, although no mention of IBT or problems with fraudulent data was made to Congress in proposing the amendments.

One of the most significant amendments permitted conditional registration of a new or already registered product before adequate testing had been completed and data gaps filled. EPA, upon discovering massive fraud and "deficiencies in health and safety data on pesticides," maneuvered Congress into changing the pesticide law to permit registrations based on that data to continue.[13] The new law also provided for additional uses of a pesticide to be allowed under the same conditional registration status.

Under another amendment proposed by the National Forest Products Association and passed by Congress in 1978, pesticides could be granted a "minor use exemption" from testing requirements, permitting them to be used, without prior testing, in areas such as the Siuslaw National Forest and other places conveniently designated "unpopulated," although they contain and surround whole communities, farms, and recreation areas. The passage of these amendments to the pesticide law meant that people could be legally exposed to substances whose long-term effects were at best unknown. Only after harm was done and proven to be caused by a pesticide could its conditional registration be challenged. Thus the law intended to protect people became a law that protected poisons.

Congress may not have been aware of the connection between the 1978 amendments and the IBT fraud, but the amendments proposed by EPA offered a tidy solution to the agency's dilemma. Caught between a powerful, not entirely scrupulous industry and an unwitting, trusting public, the EPA found a way to accommodate the giants without alerting their victims.

"The concept of conditional registration is that you don't look at all the data base when you make the decision on the conditionality," Fred T. Arnold, chief of EPA's Regulatory Analysis and Lab Audits, told industry. "The concept was to try and proceed in an orderly fashion and fill data gaps and not interfere with the ability to control pests and market pesticides [italics added].[14]

Mr. Arnold was speaking at a meeting held on October 3, 1978, at the Howard Johnson Motor Inn in Arlington, Virginia, at which EPA and Canadian officials explained the IBT audit and review process to industry representatives. There were no illusions about the importance of that meeting. "There is not another issue that cuts so broadly across the chemical industry as Industrial Bio-Test has," Mr. Arnold said.

EPA officials paid lip service to protection of the public and environment from unknown effects of chemicals with registrations based on IBT data, but it was clear that the chief concern of the agency was to accommodate industry. The Canadian official on the panel expressed greater concern for safety. He advised industry auditors to report "if any adverse effects, which would otherwise be missed, are picked up in the raw data which have not been in the final report...since the main function of both agencies dealing with this is the protection of public health." The difference in priorities between the two governments was apparent.

The Canadian government, which had already reviewed forty-five IBT studies at the time of the Arlington meeting (finding sixteen of them totally invalid), hoped to complete review of its share of IBT data by April, 1981. The U. S. set no firm deadline. Apparently EPA and the chemical industry intended to replace deficient IBT data on registered pesticides with new data provided by manufacturers to support the registrations. No registrations were suspended on the basis of falsified or shoddy IBT testing, and once replacement tests were submitted the whole IBT affair could be filed away as a slight historical embarrassment without the public ever being the wiser.

The public had little opportunity to become the wiser because of "trade secret" claims which denied public access to registration data that registrants withheld as proprietary information. Although the 1978 amendments to the pesticide law specifically provided for public disclosure of health and safety data on active ingredients of registered pesticides, manufacturers sought and obtained injunctions against the FIFRA disclosure provision on grounds that it violated their Fifth Amendment Constitutional rights. They claimed that the health and safety information was intellectual property of commercial value, and government release of the information would be unlawful taking of property without due process of law. The injunctions in the leading cases were overturned by appellate-court decisions, upheld by the Supreme Court in 1981,[15] but EPA continues vigorously to protect test data as trade secrets[16]. (The disclosure provision does not apply to inert ingredients. An "inert" ingredient is defined as any substance which does not share in the function for which the product is intended. "Inert" in this context does not mean biologically inactive. Thus, a highly toxic surfactant added to glyphosate in Roundup is classified as an "inert ingredient"[17] and information on its toxicity and chemical structure is legally withheld from public scrutiny.)

The Canadian government, meanwhile, continued to be concerned about health hazards from inadequately tested IBT chemicals. In May 1980, a Canadian reporter for the Regina (Saskatchewan) *Leader-Post* was sent by his editor to investigate a rumor that the Saskatchewan government intended to ban some ninety chemicals.

"I was an agriculture reporter," Peter von Stackelberg says. "You know, I followed grain prices and harvests, that sort of thing. I was told to check this rumor out, and I tracked down the name of David Penman, the senior health adviser with the Saskatchewan Environment Department. He told me the Saskatchewan government wasn't going to ban anything, but he kept talking about an outfit called IBT.

"'I-B-what?' I said. He gave me a short course on IBT, and said the Saskatchewan government was having a hell of a time getting any information out of the Canadian government on IBT chemicals. I made a few phone calls, and got my hands on a Canadian government memo that laid it out clearly that there was a problem with IBT, that all their rat studies were useless. The chemical companies were not cooperating in their investigation. That was the basis for my first story, in June 1980.

"Within a week and a half, the [Canadian] federal government released a list of ninety-seven chemicals. I said there's more here than meets the eye. The list did not include information on what was wrong with the tests. That was my next target-finding out what was wrong with those chemicals, and what happened in those tests.

"During the next four months, I wrote about thirty-five articles on pesticides, and the IBT story gradually unfolded. I really enjoyed it. I had come into it cold. I was educating myself and the public at the same time. There were violent verbal reactions from chemical companies, of course, accusing me of 'sensationalist media' and that sort of thing. But I never wrote anything I couldn't document. The Saskatchewan government even requested documents from me that they had been unable to obtain."[18]

Von Stackelberg's articles from June through November of 1980 – the unfolding of the IBT story – were reprinted by his newspaper in a book, *The Toxic Mist*. (See Appendix G for the full text of many of the stories published at that time.) One of the many contacts he made during this time was with the California Rural Legal Assistance law firm (CRLA). In November 1980, CRLA sent him copies of Canadian reviews of studies performed by IBT on captan, a fungicide commonly used on seeds and tree seedlings (millions of which are planted by hand in reforestation projects annually). The reviews found twelve out of thirteen tests invalid, and indicated that deliberate falsification of the data was involved. The Canadian reviews had been completed two years before, but no action had been taken on the chemical, which the raw data clearly showed to cause cancer and birth defects.

On November 4, 1980, von Stackelberg reported that nearly 250 of the first 600 IBT studies reviewed in the U.S. and Canada had been found invalid. He had learned from EPA official Dr. Diana Reisa that 99 percent of long-term studies performed by IBT were invalid. Within weeks, von Stackelberg's story was corroborated by a Chemical Regulation Reporter review of EPA's status report on IBT.

The report said that 30 out of 30 chronic rodent studies reviewed so far were found invalid. Agency policy is to consider all chronic IBT rodent studies invalid unless registrants can make a strong case to the contrary. Other review results so far:

- Carcinogenicity – 16 out of 17 (94 percent) invalid

- Teratology – 37/49 (75 percent) invalid
- Reproduction – 22/28 (79 percent) invalid
- Chronic Dog – 9/17 (53 percent) invalid
- Subacute – 71/105 (68 percent) invalid
- Neurotoxicity – 11/17 (64 percent) invalid
- Mutagenicity – 11/25 (44 percent) invalid
- Cholinesterase – 14/24 (58 percent) invalid
- Fish &Wildlife – 88/118 (75 percent) invalid
- Acutes – 146/449 (32 percent) invalid
- Residue – 11/26 (42 percent) invalid[19]

EPA, however, would not identify which tests for which chemicals were invalid. "Only the companies manufacturing the products would be notified of problems with specific chemicals," Dr. Reisa told von Stackelberg. (After granting that interview with von Stackelberg, Dr. Reisa was assigned to other duties at EPA.)

Another of von Stackelberg's contacts during this time was with Northwest Coalition for Alternatives to Pesticides (NCAP) in Eugene, Oregon. After reading a story on Monsanto's herbicide Roundup by Paul Merrell in NCAP News[20], von Stackelberg contacted him. Roundup, which had been promoted heavily for use in agriculture and forestry since the emergency suspension of 2,4,5-T, was registered almost solely on the basis of IBT data. Merrell's next article for NCAP News revealed the extent of the IBT scandal in the U.S. and included a list of some 200 pesticides tested by IBT (see Appendix D). Since that time, von Stackelberg and Merrell have worked closely together investigating the machinations of the chemical industry and regulatory agencies.

The 1978 Wall Street Journal article on IBT, mentioned earlier, had quoted an EPA official to the effect that EPA was looking into problems with eight or nine other testing laboratories at the time. In April 1981, von Stackelberg filed a Freedom of Information Act request for results of EPA audits of other testing laboratories.

"The report they sent was on many more than eight or nine labs," von Stackelberg says. "Of eighty-two labs audited, there were serious 'deficiencies' in twenty-five, and the routine destruction of laboratory reports and other documents made it impossible to audit the work of another twenty-two out of the eighty-two labs.

"The kinds of things they found were IBT all over again. Rats listed as 'dead' were also listed as having been mated at the same time. Rats were listed as having died twice. There were autopsy records for test animals that were still alive, and EPA found that tumors and other adverse effects were 'underreported'. Audits of nine additional laboratories were carried out in 1980 'for cause' – meaning complaints or discrepancies in the lab work indicated some reason to believe there were problems-but there were no results of these audits."

Of the twenty-five labs with serious "deficiencies" in testing, all tests from ten are presumed invalid. For six laboratories, decisions on presumptions of invalidity are pending, and all tests from nine other labs are referred to program management for audit. To these must be added the twenty-two labs which had destroyed their raw data, making it impossible to determine the validity of their tests. Thus, over half of the labs audited had serious problems with the validity of their testing of pesticides.

By mid-1982, EPA had not revealed what chemicals were tested by any of these laboratories, or which studies on which chemicals are invalid. The public therefore has had no way of knowing whether or not any given pesticide is currently registered on the basis of valid health and safety testing.

In fact, EPA itself does not know which data supporting registrations are valid and which are not. At EPA's request, Congress amended the pesticide law in 1978 requiring EPA to set registration standards for health-effects testing of pesticides. Under the registration standard process, EPA would review all data on each chemical currently registered, as well as on new chemicals as they are submitted for registration, identifying data gaps. The agency would then issue a standard describing the data base, including data gaps, and telling the manufacturer what new or replacement studies are required for registration of the chemical[21].

"In effect," EPA told the court in 1981, "The agency is considering that all new registrations are conditional until the data gaps identified by the registration standard are filled and the products are 'reregistered' under the standard...The registered products which were granted registrations prior to 1978 are also subject to the Registration Standard review...These products will also be subject to 'reregistration' once all the data gaps have been satisfied."[22]

Until a registration standard has been issued for any chemical, old or new, and manufacturers have submitted adequate studies to fill data gaps, the health effects of that chemical cannot be determined. Meanwhile, under conditional registration status, the product can be used, and humans may be exposed to it.

The EPA issued its first registration standard in late 1980. By early 1982, registration standards had been issued for seven pesticides. Under the Carter Administration, EPA had requested funds to complete registration standards for thirteen pesticides during 1982. Under Reagan appointee Anne Gorsuch, however, the agency requested funds adequate to complete only eight standards in 1982. At the rate of eight per year, it will take EPA fifty years to complete registration standards on the over 400 pesticide chemicals requiring them. (Completion of a standard is only the first step in the "reregistration" process. Following issuance of a standard, the manufacturer must then conduct tests – often two to five year studies – to fill data gaps. Should EPA then identify the chemical as a hazard, it must initiate cancellation or suspension proceedings, which may take an additional five to eight years. In spite of the volume of evidence against 2,4,5-T, EPA has been unable to complete such proceedings against it in twelve years.)

The entire process – conditional registration, preparation of a registration standard, filling of data gaps, and possible cancellation or suspension hearings – may well consume most of the years a product is protected under patent laws. In addition, the longer a chemical is sold under a conditional registration, the greater the economic "benefits" (to manufacturers and users) to be weighed against risks (to the public and environment) in the final agency decision.

"All this means simply that registration of a chemical with EPA says nothing at all about the safety of a pesticide at the present time," says molecular geneticist Dr. Ruth Shearer. "EPA's labeling is based entirely on the acute toxicity of the chemical and has nothing to do with the cancer hazard, genetic hazard, or birth-defects hazard. These hazards are not even considered in the labeling. Therefore, applying a pesticide according to the label does not mean applying it safely. It simply means applying it Iegally."[23]

A U.S. Senate subcommittee staff report in 1976 put it even more bluntly.

"The EPA has misled the Congress, the General Accounting Office, and the public regarding its pesticide programs," the report said. After examining the inadequacies of pesticide testing in some detail, the report accused the EPA of negligence. "Nevertheless, the EPA on many occasions gave the impression to those outside the agency that reregistration was being conducted in a thorough manner"[24]. Four years later, in 1980, the General Accounting Office reported that the reregistration program was "already behind schedule and has many unresolved policy and procedural issues which jeopardize its success"[25].

The response of the pesticide industry to Peter von Stackelberg's articles was that IBT was just an aberration, an isolated incident that didn't reflect on the testing of chemicals in general. The president of the Canadian Agricultural Chemicals Association, A.D. St. Clair, said publicly that IBT was an aberration and that he hadn't heard of any major problems at other testing laboratories in the U.S.[26]. St. Clair had until recently been a vice president of Diamond Shamrock, owner of a lab where an EPA investigation uncovered so many problems that the agency took the position that "studies performed by this firm are unacceptable for regulatory purposes."[27]

"I had been offered a job in Edmonton, Alberta," von Stackelberg says, "And this was my parting shot in Regina. I called St. Clair and confronted him with the audits on those eighty two labs, including the one owned by Diamond Shamrock. There was a long silence on the other end of the phone. I started reading off the reports on the other labs. His comeback finally was, 'I'm shocked. We knew nothing at all about this.'

"So things hit the fan all over again. Penman was barraging the [Canadian] federal government with questions, and the credibility of the chemical industry was going rapidly downhill. It was IBT all over again – only twenty-five times over.

"The Canadian government claims it didn't even know about the other labs. The trouble is, Canada relies entirely on test data from the U.S. given to EPA There are no commercial or corporate labs for testing pesticides in Canada, and damn few for drugs. So when all this came out, Canada felt badly shafted by the U.S.

"Industry has accused me of being anti-technology, being out to destroy the chemical industry. But I got into this thing with no feelings one way or the other. I had no bones to pick with the chemical industry. I didn't know anything about it. The kinds of outrageous statements industry made set me off. I just couldn't buy their thing, and went out to find out. After I started seeing what was going on, after listening to their bull, after I started to look at the statistics and saw the casualties – then I got pissed off.

"They call exposing their lies sensationalist and biased. But I was just digging up what was there, telling it to anyone who would listen. They never have gotten me for any inaccuracies. You don't have to tell people what to think, just give them the facts and let them draw their own conclusions. That's all I'm doing."

* * *

In June 1981, a federal grand jury in Chicago indicted four former IBT officials on charges of falsifying test results on chemicals and drugs between 1969 and 1976: Jjoseph c. Calandra, former president and founder; Moreno L. Keplinger, former manager of toxicology; James B. Plank, former assistant manager of toxicology; and Paul L. Wright, former section head of rat toxicology. (Wright, who held that position at IBT from March 1971 until September 1972, has worked for Monsanto since October of 1972 as manager of toxicology in Monsanto's department of medicine and environmental health[28]. Registration of Monsanto's herbicide Roundup was based almost entirely on IBT data.)

The four men were charged with falsifying reports to the FDA and EPA, in particular with fabricating results of two cancer studies on the herbicide Sencor and the insecticide Nemacur (manufactured by Chem-Agro, a subsidiary of Mobay Chemical Corporation).

According to the indictment, the officials concealed the fact that TCC (trichlorocarbanilide), a Monsanto product used in deodorant soaps, caused atrophy of the testes at the lowest dose tested in mice[29]. They also fabricated results of blood and urine studies that were never performed on Syntex Corporation's antiarthritic drug Naprosyn, the indictment charged.

Naprosyn, TCC, Nemacur, and Sencor are still registered and on the market. In late October 1981, the Canadian government recommended banning the use and sale of six pesticides registered on the basis of IBT testing and recommended cautionary statements to be added to the labels of thirty-four pesticides, including Roundup[30]. EPA has yet to limit use of a single chemical tested by IBT.

Of the products tested by other laboratories we know nothing, and EPA is tight-lipped. Within the government itself, there is considerable doubt about the quality and integrity of EPA reviews of toxicology data. Dr. Adrian Gross, a veterinary pathologist with FDA, found at least one EPA evaluation of a study on the pesticide Permethrin to have deficiencies as serious as the deficiencies in the study itself. Calling the EPA review "egregious," Gross conducted a personal audit of the EPA evaluation. His 72-page review demonstrates that EPA had "underestimated the risk by three to four orders of magnitude[31]. He concluded that very important problems with the experimental study had not been perceived by the EPA reviewers, and that the pesticide was many times more carcinogenic (cancer-causing) than was indicated by EPA's 'superficial and flawed' evaluation. In risk estimates for liver tumors, Dr. Gross found that both the registrants and EPA reviewers had included statistics on a large number of animals which had died before the possible onset of such tumors. To consider these dead animals 'at risk' for liver tumors, Dr. Gross concludes, would be like "believing that the best way to avoid contracting cancer in one's old age is to step in front of a moving bus while one is still young."

On the basis of this flawed evaluation, permethrin was approved under an "Emergency Exemption" for use on a very large scale in field corn in several states in 1980. The experimental study which Gross found to have "serious faults" was conducted by Biodynamics, Inc., a private testing laboratory which had been audited by EPA and approved under a "No Action Indicated" status[32]. Dr. Gross's scathing review casts doubt both on the integrity of the laboratory's work and on the reliability of EPA audits.

Unknowns about the effects of particular chemicals are compounded by even greater unknowns about the combined, additive, or magnified (synergistic) effects of chemicals in combination. The billion pounds of pesticides used annually[33] in the U.S. are concentrated on the nation's food and fiber crops.

"The truth is that each American citizen now carries some individual burden of pesticide in his bodily tissues," Dr. Doutt says[34]. "If EPA knows what the long-term effects of those pesticides are, either separately or in concert with other pesticides, drugs, industrial chemicals, nuclear radiation, or other toxins, the agency is doing its best to keep the public from finding out."[35]

EPA is also not willing to tell the public where the fraud lies. Deputy Assistant Administrator Edwin L. Johnson acknowledged that "evidence is accumulating which suggests prior knowledge of those practices [falsifying data] by the sponsors" of health and safety studies on pesticides. In the same internal EPA memo, Johnson wrote, "I know that you share my deep concern regarding the seriousness of the regulatory ramifications of these recent findings of falsification of data upon which national and international regulatory decisions have been based"[36]. Whether this deep concern was for the public or for his own reputation, Johnson does not say.

Regulation of pesticides has wittingly or unwittingly become a cruel hoax played on a trusting public. In the remote, unpolluted landscapes of our national forests, human beings have involuntarily become test animals for federal agencies' experiments with fraudulently registered poisons.

The Five Rivers Health Study

The major problems of environmental pollution were not predicted in the laboratory. Bound in the straitjacket of positivism, many scientists refuse to acknowledge the warnings posted by circumstantial or correlative evidence, waiting instead for the "scientific proof" provided only by the establishment of cause-and-effect relationships.

—Frank Graham

While science, industry, and government have largely ignored or dismissed the circumstantial evidence of pesticide effects offered by exposed populations, they have continued to make decisions based on laboratory data known to be invalid. Pesticide chemicals are thus presumed innocent until dramatic evidence of human harm proves otherwise. Even such dramatic evidence as the Alsea study was only sufficient for the token "suspension" of some uses of 2,4,5-T issued by EPA. Subsequent hearings resulted in secret negotiations between EPA and Dow Chemical Company, which effectively prevented the completion – or public disclosure – of risk/benefit analysis. Further evidence of human harm from herbicides studied in the Siuslaw National Forest from 1979 to 1981 has met with an equally disturbing fate.

After the emergency suspension of 2,4,5-T, the Forest Service substituted 2,4-D and picloram in its Siuslaw National Forest spring applications in 1979. Although they applauded the 2,4,5-T suspension, Five Rivers residents were distressed about the spring spraying. They were well aware that 2,4-D had been sprayed along with 2,4,5-T and silvex in the Alsea study area, which included Five Rivers. EPA had never offered a reasonable explanation why only dioxin, and not the herbicides, had been implicated in the miscarriages and other health effects reported. The Forest Service proposed to spray units within a mile or closer to private homes in 1979, including the watersheds for several domestic water supplies.

Five Rivers residents presented their concerns to the Forest Service at a public meeting with District Ranger Ned Davis in May. They reminded him that until the EPA suspension of 2,4,5-T and silvex, the Forest Service had adamantly maintained that the chemicals were safe. How could the public trust similar statements the Forest Service made about other chemicals? they asked.[1] Davis was noncommittal in his replies to people's questions, but he did express concern about individual water supplies. He noted the locations of all water sources within proposed spray areas, as pointed out to him on a map by local residents, and assured them that spray plans would be revised to avoid contaminating their water.

Four days later, however, when the spraying began, the watersheds pointed out to Davis were sprayed, along with all the other proposed units. Within weeks of the spraying, three of the five pregnant women in the valley miscarried.[2] In July, the three women met with neighbors who had also experienced health problems following the spraying and drafted a letter, which they sent to EPA, the Forest Service, and the Lincoln County Health Department:

Dear Sirs:

We are residents of Five Rivers, a community of about 50 families in the Alsea drainage of the coast range in western Oregon.

This spring, 1979, the U.S. Forest Service sprayed 742 acres of forest in and around our valley with 2,4-D and Tordon 101 (a mixture of 2,4-D and picloram). The 2,4-D was most widely and heavily used, applied at the rate of 10 gallons per acre on 612 acres; the 2,4-D and picloram mixture was applied at the rate of 20 gallons per acre on 130 acres. The spraying was done between May 12 and June 5. Between May 25 and July 12, three of the only five women known to be pregnant in the valley had spontaneous abortions. All three women were treated by physicians and were in the first trimester of pregnancy. The two surviving pregnancies were in later stages of gestation. The first miscarriage occurred on May 25, the second on May 27, and the third on July 2. All three women live within one mile of heavily sprayed units.

Since May 12, the health of the population of our valley has undergone profound and disturbing changes. Chief among them has been the incidence of respiratory illness and intestinal disorders in almost every household. Several women experienced uterine hemorrhaging not associated with pregnancy. Children and adults suffered bleeding gums, bloody noses, and a number of women suffered from bacterial vaginal infections. Ten children and three adults have experienced an undiagnosed illness characterized chiefly by high fever; two children and two adults had undiagnosed rashes that doctors were unable to explain, and on July 9 occurred one near-fatal case of meningococcal meningitis in a two-year-old who is still hospitalized.

This is by no means an exhaustive survey of our community's health, only a summary of health problems in the families of our acquaintance. In a population of our size, such an incidence of illness in what was previously a healthy population, occurring within so short a time, is most alarming, especially to the people who live here. It is difficult . . . to attribute it solely to coincidence, and the only stress the population has experienced as a whole has been exposure to the herbicides sprayed.

We urgently request that these herbicides cease to be sprayed until their safety has been unequivocally established, and that the miscarriages and the health of the population be studied immediately. A number of families are severely in debt as a result of the deterioration of their health since the spraying, and we can ill afford the problems that recur each time these herbicides are applied. Time is of the essence. The price of our present national forest policy is our children's lives[3].

The women accompanied their letter with a chart comparing the dates, locations, acres sprayed, herbicides applied, and dates of miscarriages. The response of the Lincoln County Health Department was immediate and gratifying. On July 16,[4] the county initiated a health study of the Five Rivers area, meeting with residents and composing a health questionnaire for them to complete. In response to fears about the sudden case of spinal meningitis, the health department also collected blood and urine samples and throat cultures from any residents who wished to be tested.

By mid-August, a second child had been hospitalized with another near-fatal case of meningitis. The Forest Service announced that it was about to begin its fall spray program, using Krenite and Roundup. No reply had yet come from EPA to the women's letter. On August 14, they wrote a second letter to EPA Administrator Douglas Costle, repeating their concerns about health and expressing their alarm about the fall spray program using herbicides "about which even less is known than about 2,4-D or picloram."[5]

"We appreciate the fact that EPA is now actively engaged in studying retrospectively the health effects of 2,4,5-T and silvex" the women wrote. "This is a necessary and important task. But we are experiencing effects from other registered chemicals, and cannot wait another five years for EPA to protect us from these...EPA must act immediately to protect us from currently registered chemicals which pose unknown risks to our own and future generations."

Within a week of mailing their second letter, the women received from EPA a reply to their first letter. The letter informed them that EPA, the Oregon State Health Department, Lincoln County Health Department, and the University of Colorado Epidemiologic Studies Program (ESP) were conducting a cooperative study of the problems in Five Rivers. Colorado ESP had been contracted by EPA to perform epidemiological studies for the previous Alsea study[6]. About the time the EPA letter arrived, E.S.P. researchers visited Five Rivers, interviewed the women who had miscarried, and collected water and sediment samples from local water supplies. According to the EPA letter, the samples would be analyzed for residues of 2,4-D, picloram, 2,4,5-T, silvex, and TCDD.

Although willing to conduct such studies, EPA showed no apparent inclination to persuade the Forest Service to curtail its fall spray program. On August 29, Dr. Barbara Wood, the Lincoln County health officer, wrote a letter to the county commissioners, asking them to request the Forest Service to delay their program. Having studied questionnaires from Five Rivers residents, she wrote:

> Preliminary review indicates....a definite increase in respiratory and gastro-intestinal problems, as well as an increase in vaginal bleeding, with three miscarriages reported, following the spraying of 2,4-D in the latter part of May.
>
> I would propose that an in-depth herbicide epidemiological study be done in this area and throughout Lincoln County, because at present, there is no assurance that herbicides are not harmful to health.
>
> I request the Lincoln County Board of County Commissioners to consider requesting the U.S. Forest Service...to delay the scheduled aerial spraying of the herbicides "Roundup" and Krenite for the present, and further request they initiate a plan to study the incidence and effect of herbicide use.[7]

The following day, the Lincoln County Board of Commissioners wrote to the U.S. Forest Service, enclosing Dr. Wood's letter, and requested a delay in the fall spray program[8]. Siuslaw Forest Supervisor Larry Fellows refused to comply with this request, claiming that the Lincoln County study was concerned with phenoxy herbicides which were not being used in the fall program. He further noted that the Forest Service was cooperating with the Colorado ESP and EPA in their research "to review and monitor the fall application program."[9] Neither the Five Rivers people nor their county officials were able to stop the fall spraying, and being assured that the effects of the herbicides on their health would be "reviewed and monitored" was small comfort. (According to an October 4, 1979 EPA status report, "it is planned to continue work with these agencies by assisting in the administration of the second questionnaire in November after the end of the current spray season. Specifically we will be looking for the occurrence of pregnancies and spontaneous abortions, as well as the onset of health effects as may develop following spraying of Roundup and Krenite".)[10]

On November 2, 1979, after the Forest Service had completed its fall spray program, EPA Deputy Administrator for Pesticide Programs Edwin L. Johnson answered the second letter from the Five Rivers women. In a lengthy letter to Mrs. Melyce Connelly, he outlined the health study being conducted by EPA's Colorado Epidemiologic Studies Program and the local county health department, and described the analyses being performed on local samples for 2,4-D, picloram, 2,4,5-T, silvex, and dioxin.

"Through this study of the health problems you have alerted us to in the Five Rivers area, we hope to learn whether routine, seemingly lawful pesticide use is resulting in hazardous exposure of local residents to pesticides," Johnson wrote. He described EPA's process of balancing risks against benefits in its registration of pesticides, suggesting that the Five Rivers women were seeking "different, broader pesticide regulatory remedies than...EPA is authorized by law to provide."

"Please understand," Johnson concluded, "that while the law requires that the 'burden of proof' for demonstrating that a pesticide's risks are reasonable remain with the registrant [manufacturer], it also requires EPA to have a reasonable basis for initiating action to take a pesticide off the market. Thus, hard evidence of unreasonable risk is needed for EPA to take suspension or cancellation action against any registered pesticide produce."[11]

"In other words," Mrs. Connelly said, "he's saying it's out of their hands, that once something is registered they can't unregister it. EPA trusts the makers of the chemicals to tell them about 'risks', but not the people who get poisoned. They're not going to do anything without 'hard evidence' – meaning we'll all have to be dead and analyzed before they're going to believe us, or take any action."

Ed Johnson, who wrote the letter to Mrs. Connelly, had "taken a very active role in [EPA's] Lab Audit Program and, particularly, in questions dealing with Industrial Bio-Test [laboratories]."(See Chapter 12)[12]Through his involvement with the IBT lab fraud, he was certainly aware that the concerns of the Five Rivers women could indeed be valid, and that in fact EPA did not have adequate knowledge of the effects of pesticides to judge the reports from exposed forest residents. Picloram and 2,4-D-applied in the spring forest spraying were among the chemicals tested by IBT, as were herbicides Roundup and Krenite, which the women begged EPA to prevent the Forest Service from spraying in the fall.

In effect, what Johnson's letter proposed was that EPA's study of health problems in Five Rivers could provide the "hard evidence of unreasonable risk" needed for EPA to take action. In a speech to the Forest Industry's Environmental Forum in Washington, D.C., a copy of which he attached to his letter, Johnson had emphasized the importance of animal testing to estimate human risk from pesticide exposure.

"Prudent administration does not, indeed cannot, wait until an effect is observed in a human population," he said. Evidence of human effects "must be considered a measure of program failure, not program success, since we should have dealt with the issues before they became widely evident in a human population."[13]

The multiagency health study of the Five Rivers area was apparently such a measure of "program failure" in light of what EPA and Johnson knew about IBT and other testing labs at the time he wrote his letter to Mrs. Connelly. Within a week after writing that letter, Johnson circulated his Lab Quality Status Report within EPA, revealing grave problems with testing in over half the laboratories audited[14]. The previous year, on December 5, 1978, Johnson had published a Federal Register notice establishing tolerances for residues of the herbicide Roundup on a variety of raw agricultural products. According to Johnson's notice, the data submitted by Monsanto had been evaluated by EPA, and in spite of numerous "desirable studies that are lacking or to be repeated" – including birth defects, cancer, and mutation studies – the tolerances were approved by EPA[15].

What is interesting about the granting of the tolerances is that all of the studies listed in the Federal Register notice were performed by IBT. According to documents released by the Canadian government in 1981, nearly all of those studies were found to be invalid[16]. If, as Johnson states in the notice, those studies had been "evaluated" by EPA at the time (1978), then EPA was well aware that tolerances for Roundup on food were established on the basis of invalid or nonexistent animal tests.

Testing inconsistencies with herbicides 2,4-D and picloram were outlined at a symposium on pesticides in Portland, Oregon, in July 1979 by Dr. Melvin Reuber of the National Cancer Institute. Dr. Reuber had been asked by a Senate subcommittee to review the test data for 2,4-D and picloram submitted by Dow in support of registration. He found that the raw data on 2,4-D did not support Dow's conclusions that the chemical could not cause cancer, and that no raw data existed to support Dow's claim that picloram did not pose a similar risk[17].

EPA – or at the very least Deputy Administrator Edwin Johnson – was fully aware of the magnitude of these testing "deficiencies" when forest residents reported the events in Five Rivers in 1979 and appealed to the agency for help. Understandably, the agency was interested in studying Five Rivers health problems. It was an opportunity to learn whether, indeed, as the Federal Register notice proclaimed, EPA's regulatory procedures "will protect the public health."

The people of Five Rivers did not know of the IBT scandal in 1979. What Johnson's letter meant to them was simply that EPA was refusing to act to protect them from Forest Service use of poisons. A press release from the forest supervisor suggested ominously that little enough was known about Roundug and Krenite to justify "monitoring" their effects and efficacy[18]. When the county government failed to convince the Forest Service to delay the spraying, there was no more anyone could do. The Five Rivers Health Study continued, and people cooperated as best they could. In September, a four-eyed kitten was born in a sprayed area of the valley. The Colorado researchers took the kitten for analysis, along with water and sediment samples, promising results as quickly as possible[19].

The Five Rivers women were unaware, however, of the scope of the EPA study they participated in, which included hospital and physicians' records from the area as well as the collection of other samples for analysis. The study was already in progress at the time the women wrote their first letter, reporting health problems in the valley. EPA documents released in 1981[20] indicate that the study was intended as a follow-up to the Alsea study, to determine whether the health effects assumed to be linked to 2,4,5-T and dioxin would be diminished since the emergency suspension of that herbicide. Accordingly, EPA researchers were on the scene in the area when the Forest Service sprayed 2,4-D and picloram that spring, ready to collect samples and monitor health effects – particularly miscarriages – as recorded by hospitals and physicians. What later became known as the Five Rivers Health Study actually began before the women wrote their letter and was designed to be a prospective study continuing at least through 1980.

The few documents EPA has released reveal that the Five Rivers Health Study was a major epidemiological study, using the Alsea II study as a base to which future health effects would be compared, with the birth records of the entire state of Oregon as a control. The study involved both 2,4,5-T and silvex and the chemicals which replaced them after the suspension. At least one of those replacement chemicals – Roundup – was being sprayed under an "experimental use" permit. The results of the study could have a profound effect on the outcome of the 2,4,5-T cancellation hearings, as well as the Agent Orange veterans' class action suit. If the study showed a decrease in health problems after the ban on 2,4,5-T, registration of this economically insignificant herbicide could be cancelled, satisfying public concern and improving the tarnished image of EPA. Use of more profitable replacement chemicals could escalate unabated.

The letter from the Five Rivers women, however, suggested that health problems continued, and perhaps increased, following the spraying of those replacement chemicals – all of which had been registered without EPA evaluating their safety tests. If the study results supported this possibility, some of the most profitable products in the chemical industry – notably 2,4-D and Roundup[21] – would be affected. Five Rivers residents also learned from the documents released in 1981 that in addition to the four-eyed kitten and the water and sediment samples taken by EPA researchers in 1979, small birds and mammals had been trapped in the area, and tissue samples had been taken from an anencephalic (without a brain) baby born that year. The researchers had also taken for analysis a miscarried fetus, placenta, and other "products of conception" from a local woman.

"They took it – the baby – away in a plastic bag," the woman recalls. "They promised to let me know the results of the testing as soon as possible, and put me through the third degree with this questionnaire they had. That was almost three years ago. I've never heard a word from them since. The watershed for our whole community was sprayed that year, and there was a lot of controversy over it. We never could find out who sprayed, or what they were using. Then they came and took samples, took the baby to be tested, and never told us what they found.

"So all this time they leave us wondering what's in our bodies, what's in our children's bodies that they can't live long enough to be born. My neighbor and half the women I know have had to have hysterectomies. The whole thing is just sick. You never hear of any women flying around in helicopters spraying women and children with poisons..."

From July through December, 1979, in the small community of Ashford, Washington, within an area that had been heavily sprayed with 2,4-D, only two live births occurred out of twelve pregnancies conceived during those months. Of the two live births, one infant died within two weeks of a congenital heart defect. At a public meeting of worried, grieving residents, Dr. Ilio Gauditz, a consultant for Weyerhauser (the timber company responsible for most of the spraying), dismissed the women's concerns.

"Babies are replaceable," she told them, and advised them to time their pregnancies not to coincide with the spraying[22].

Neither the women of Ashford nor the community of Five Rivers shared Dr. Gauditz's cavalier attitude toward their children. In the summer of 1981, as the Forest Service prepared to spray the forests surrounding Five Rivers with Roundup and Krenite, residents of the valley petitioned Senator Mark Hatfield, asking him to make inquiries about the Five Rivers Health Study.

From early 1979 to late spring of 1981, they told Hatfield, only one live birth had occurred in Five Rivers[23]. EPA's response to Hatfield's inquiry was that the study had not been completed and no results were available[24].

* * *

During the summer of 1979, two physicians from the northern part of Lincoln County in the Siuslaw Forest attended the symposium in Portland where Dr. Reuber presented his testimony on 2,4-D and picloram. Doctors Renee and Chuck Stringham were family practitioners from Lincoln City. Their first acquaintance with the health effects of herbicides had occurred two years earlier, when some residents of nearby Rose Lodge, who had been sprayed by Publishers Paper Company as they occupied the company's land to protest its use of herbicides near their farms and watersheds, came to them for treatment of various symptoms. Blood and urine samples from ill patients showed measurable levels of herbicides.

Following that experience, Renee and Chuck Stringham began questioning other patients with similar symptoms about possible exposure to the chemicals. They also became concerned about the unusual incidence within their own practice of malformed infants born in the area, in particular types of birth defects affecting the brain and spinal cord[25].

During their four years of practice in Lincoln County, the Stringhams had delivered 300 babies. Nineteen of them were born with birth defects, seven of them neural-tube defects. Two of these seven were anencephalic. Such infants are usually healthy and normal in appearance, except that the skull ends in a bony ridge above the eyebrows with a thin veil of tissue covering an empty brain cavity. Anencephalic infants can be kept alive for a few hours or days at most. One of the samples collected by the Colorado Epidemiological Studies Program team for EPA was tissue from an anencephalic infant born in the Five Rivers study area in October 1979[26].

(Two out of 300 births is more than thirteen times the national average of five anencephalic infants for every 10,000 live births. The incidence of neural-tube defects in their patients was five times the national average. Neural-tube (brain and spinal cord) defects, resulting in central-nervous-system damage, are birth defects related to damage to the neural tube during its embryonic development. They include anencephaly (absence of a brain); hydrocephaly (abnormal accumulation of cerebra-spinal fluid within the skull, causing enlargement of the head, atrophy of the brain, mental deterioration and convulsions); porencephaly (abnormal cysts or cavities in the brain); encephalocele (protrusion of brain substance through a congenital abnormal opening in the skull); and spina bifida (incomplete development of the spinal cord, involving leakage of spinal fluids, paralysis, and brain damage).)

Within that study area is the small town of Waldport, at the mouth of the Alsea River. Waldport is primarily a service community, serving tourists, sport fishermen, timber workers, and the scattered population of the Alsea River valley and its tributaries, including Five Rivers. Along the lower Alsea, up to the limit of the tidewater, are clustered small trailer courts and retirement and vacation homes, distinguished by the tidy docks, drift boats, and other small craft used by a devoted community of steelhead fishermen.

In a small travel trailer in one of these trailer courts lives Larry Archer, with his wife and their two-and-a-half year old daughter, Kaleen. Larry works at a gas station in Waldport. In October 1979, his wife Laura gave birth to a child with anencephaly. Larry is a young man in his early twenties, wiry and strong. In the cramped, stuffy trailer on an overcast spring evening in 1980, he wears a T-shirt with its short sleeves rolled up to his shoulders. He sits in the hallway on a cushioned bench that doubles as a child's bed, holding his sleeping daughter Kaleen on his lap. In the kitchen a few feet away, Laura silently prepares dinner.

"You want to know about the baby?" Larry says. "What do you need to know? It's dead. You don't want to know the rest. But I can tell you about sprays. There's a lot more I know about sprays than you'll find out from those ass-holes at the university. Some fuckin' nerve they've got. Did they ever see a baby like that? Did they ever see a baby like that and know it was theirs – their own kid? Just let one of 'em tell me it couldn't be the herbicides. I can't prove nothin', but I know."

His voice rises in anger.

"Look, I lived around here all my life. Up in Siletz, that's where I grew up.

They used to spray there every god damn year. We was just little kids then. We thought it was so neat – they loaded the helicopter right at the end of our driveway. We'd all run out there – running around the helicopter, getting in the way while they loaded it. It was such a trip, you know, having a real helicopter come down right there... I mean, we didn't know – and those guys never said a word about the stuff, that it wasn't good stuff to get on you. It'd be dripping all over us. And then we always ran after the helicopter when it took off. We was kids – it was a trip, and it'd be spraying the stuff all over us with us running after. They never warned us, never told my folks nothin'. Twelve years they did that every year.

"My folks bought our milk from a neighbor who sprayed the pastures with the cows still out on 'em. They never thought nothin', thought it was that safe. So we was probably drinkin' the stuff all that time, too.

"So I grew up with that stuff. They sprayed it all over the goddamn place. Who knows what I got in me, for chrissake? Laura, she comes from L.A. I don't know, but they probably don't spray right in the city there. Except we was living in Corvallis before we came back here – while she was pregnant last time – and they was even spraying there, right up by the god damn reservoir. We moved back here before our second baby was born, though.

"Everything was going real good then. Laura was fine. It was a good pregnancy, I mean – right up to the end. Then, two days before the baby was born, the doctor says there's something wrong, he can't feel the head. So we go over for xrays. And I'm sittin' there, waiting for Laura, you know, and the doctor comes out with this bad look on his face. He says, 'There's something I have to tell you. I hate to tell you this, but...' And he goes on, saying there's something wrong with our baby. Something's wrong with its head, and he says it probably won't live. So we had two days to think about that, and then Laura goes in labor, and all the time we're knowing this."

Kaleen sleeps peacefully on Larry's lap, his hand stroking her hair.

"We went in the hospital together. When Kaleen was born, you know, I was there the whole time, watching. And maybe this sounds corny, but that was the most beautiful moment of my entire life – just being there...

"So I was there this time, too, but it was another thing altogether. The baby – it was a girl. She was perfect, from her toes up to her eyebrows. I mean her face was perfect, too – kind of like Kaleen's, almost. But that was it. It ended at the eyebrows. That's all there was – just this kind of bowl, with a kind of film of tissue over it. She couldn't breathe. There wasn't any brain to tell her to breathe. So they give her oxygen, and she lived for about an hour. An hour and twelve minutes. I don't know how Laura stood it. I would've broken up then, but she didn't.

"These people who say herbicides are so safe – I could give them something to think about. I'd like to put everyone of 'em in a delivery room to see a baby born like that, like 'em to have to watch their own kid born like that. To them, to everyone, our baby was just a statistic, a number. I want 'em to see the real thing. They just want numbers. They want to blame it all on anything but what it could be. And they sure as shit don't want to find out.

"I made the doctor take samples – tissue samples – from her, from the baby. He was supposed to get them tested-at OSU, he said, and another one at Colorado. That was a whole year ago [1979], and we never heard nothin' about it. I went in there once to ask him about it, and he wouldn't even come out and see me. He was there, and he wouldn't even come out and talk to me. That's how much he cares, how much he wants to know.

"I knew, see. I knew how much of that stuff I got on me all the time I was growing up. My brother and me – we used to spend all our time in the woods around our place – I mean all our time. That's where we went when there was nothing else to do. And we used to find little deer embryos – about the size of a cat they were, all in their little sac – whole – just dropped where they fell out of the mother. Three of 'em we found one year. We thought they were babies – human babies – at first. It was real spooky, till we looked close at one and could see it was a deer. We never thought nothin' about 'em then, except they were so spooky, just findin' 'em layin' on the ground like that.

"So when our baby was born, I knew. I asked the doctor could it be me – could it be that stuff in me? And he said I could go up to Portland and have my blood tested and all, but it would cost – oh, six hundred dollars a test or something. Fat chance of that. I work in the gas station, and we have this trailer, and we pay the rent for the space, and I'm still paying the hospital bill and funeral bill for our baby...

"It's been a whole year," Larry says, "But you don't get over it, because you're afraid. Afraid to have another baby, afraid of what maybe is already in the one we have, of what will happen when she grows up and has kids. And no one will listen to you, and there's not anyone – not even the ones who love the stuff so much – can tell you it's all right. They just want their money, their profits, and we're just a number to them.

"You grow up thinking the law is there to protect you, and then you find out it's just there to protect the people with money. That's how much good their laws are. I'm not going to go out and hurt anybody, but fuck their laws. It's their laws that let them spray this shit all over – all over the roads and the berries and the deer, and into people's water and the streams and rivers, the air we breathe. And the laws that let 'em do that don't count us one bit.

"What can anyone do? How bad do things have to get before anyone will listen? Do any politicians care about anything except bombs and nuclear plants and money? Do they care about the trees and the fish – the whole ocean – dying? Or babies that aren't even born, that won't have a chance to live?

"The Bible says the Lord will bring the end of the world. Bullshit. People are doing it all by themselves. They don't need the Lord's help. If you care, you can't sit back and watch. You do what you can. I don't know what I can do except tell people what's happening—what's happened to us."

* * *

The Stringhams, of course, could not prove a correlation between herbicide spraying in Lincoln County and the birth defects they were seeing, but other research indicated a plausible connection[27]. A similar increase in neural-tube birth defects had been noted in sprayed areas of Vietnam[28] and such problems were frequently reported in children of U.S. Vietnam veterans[29].

Along with thirteen other physicians from Lincoln City, the Stringhams petitioned the Lincoln County commissioners on September 13, 1979, to institute a moratorium on the use of phenoxy herbicides in Lincoln County "until the pesticide users and producers can prove beyond a reasonable doubt that these chemicals are safe for human exposure[30]. Several months later, ten of the twelve physicians on the Newport hospital staff signed a similar petition to the commissioners[31]. (Lincoln County's two hospitals are located in Lincoln City and Newport, and the petitioners were all but two of the county medical society's doctors.)

The commissioners refused to act on the petitions, apparently in the belief that the county would have no jurisdiction over state and federal activities within the county[32]. When a statewide initiative petition to restrict phenoxy herbicide use (supported by physicians from Lincoln County and other heavily sprayed areas of Oregon) failed to obtain the required number of signatures, the Lincoln County Medical Society sponsored two herbicide measures on the Lincoln County ballot in November 1980. The measures would have imposed restrictions on roadside and aerial applications of herbicides within the county, designed to protect watersheds and minimize contamination of private lands and community water supplies[33].

The Lincoln County initiatives attracted a massive publicity campaign funded by multinational corporations and out-of-county special-interest groups[34]. Expensive literature was mailed to registered voters, misrepresenting the ballot measures, predicting jail sentences and fines for citizens treating dandelions on their lawns. A spurious organization calling itself Doctors for Facts, none of whom lived or practiced in Lincoln County (at least one of the doctors listed on its letterhead had not given permission for use of his name), also mailed campaign literature to voters, presenting wild claims of herbicide safety as "scientific" facts[35]. The campaign material relied heavily on such scientific experts from Oregon State University as Frank Dost and James Witt[36].

"I don't know what it is with these guys, the scientist-advocates of pesticides," says biologist Dr. Steven Herman of Evergreen State College in Washington. "I don't think they're into it for the money. I think they become trapped in their own rhetoric. They become evangelists, on some political-religious crusade to defend these chemicals and evangelize about them. They're always very self-righteous. You can't find very many of them who has ever said, 'Sorry, I was wrong.' They label anyone who doesn't agree with them 'emotional, unscientific, unprofessional, immature' but they themselves are very heavy on belief. They honestly believe they and nobody else is right. I don't think money has much to do with it, but they all end up tied to industry, which is just a bunch of highly organized rednecks"[37].

Industry spent at least $67,000 to convince Lincoln County voters to defeat the ballot measures[38]. People might have voted differently had they known in November 1980 what EPA knew in 1977. The Forest Service had done little spraying in the Siuslaw during 1980, and county road crews had honored "NO SPRAY" signs along roadsides. Some voters perhaps were lulled by the respite from immediate herbicide activity. Months later, NCAP News broke the story of the IBT scandal, including a list of pesticides tested by IBT to support registration[39]. On the list were all the herbicides applied in the Siuslaw National Forest by the U.S. Forest Service.

* * *

In January 1981, the Forest Service announced its plans to triple the acreage to be sprayed, timing its local notice to provide Siuslaw residents only four days to respond and comment. Over nine thousand acres of the Siuslaw were scheduled to be sprayed, with heavy concentrations of proposed spray units surrounding the Five Rivers valley. The chemicals to be used were the same applied in 1979 – 2,4-D and picloram in the spring, and Roundup and Krenite in the fall.

Public response to the Forest Service spray plan was vociferous. Forest residents and workers protested not only the plan but the four-day time limit for public comment. At a Five Rivers meeting, which the district ranger refused to attend, residents drafted a letter of protest, requesting a 30-day extension to the comment period. Their criticisms centered on the fact that the Forest Service proposed to apply herbicide poisons without an environmental impact statement describing the chemicals' effects and the possible alternatives to their use – a requirement firmly established by the court in the CATS case, brought by valley residents in 1976. Instead, Forest Supervisor Larry Fellows had published an environmental assessment report, which included no information about the herbicides. The report declared that the spray project would have "no significant impact" on human health or the environment and therefore did not require an EIS[40]. With this arbitrary claim, the supervisor absolved the Forest Service of its legal obligation to study – and disclose – the effects of its actions.

On February 14, 1981, Paul Merrell petitioned the administrators of EPA, the Forest Service, and the Bureau of Land Management "to jointly prepare an environmental statement on the proposed [herbicide spraying]." He argued that both the Forest Service and BLM sprayed herbicides in the general area of the Siuslaw, and that both evaded the requirement of an Environmental Impact Statement by abdicating to EPA registration as assurance of public safety. Therefore, Merrell argued, a reasonable assessment of herbicide impacts could only be made by all three agencies working together, since only EPA knew whether or not the safety testing on forest herbicides was valid.

When his administrative appeals were denied, Merrell appeared in Federal District Court in Eugene to sue the three agencies (EPA, the Forest Service, and BLM) for a jointly prepared EIS which would adequately address the "deficiencies" in health and safety testing evidenced by the IBT scandal, and requested the court to issue an injunction against the spraying until essential health and safety studies were completed[41]. He appeared in court without an attorney, and was permitted to proceed in forma pauperis (in the form of a pauper), sparing him the expense of court fees. He sued also for a temporary restraining order to prevent the Forest Service from awarding contracts until the case was decided.

Federal District Judge Robert C. Belloni denied the request for a temporary restraining order because Merrell had not proved that irreparable harm would occur. The following day, however, the Alsea District, which includes Five Rivers, announced cancellation of all spring aerial spraying in the district, and a substantially reduced plan for late summer spraying. In 1982, Merrell's case was still pending, repeatedly delayed by the government's refusal to provide the plaintiff with evidence relating to the Five Rivers Health Study, EPA evaluations of safety testing on forest chemicals, or connections between testing laboratories and the military chemical and biological warfare command[42].

Merrell's suit had the immediate effect of curtailing Forest Service spraying in Five Rivers in 1981, and spraying in the rest of the Siuslaw was reduced, ostensibly in response to public concern. It turned out, however, that most of the 3,000 acres cut from the spray program had not even been logged yet. Apparently, those acres were deliberately included in the proposal with the intention of dropping them in the "public interest."

"One Little Helicopter"

The community of Five Rivers breathed easier in the spring of 1981, temporarily freed from the threat of helicopters, poison, miscarriages and illness. Their respite was short-lived, however.

On May 12, a county spray truck drove up the valley. It was driven by James Cassens, one of the men who in 1975 had sprayed my children as they were fishing by the river[1]. Along the stretch of river where those children had been hosed with 2,4,5-T six years earlier, a "NO SPRAY" sign was posted, plainly visible. The same sign had been honored the year before by county spray crews[2].

James Cassens saw the sign. But this year, he told a reporter later, orders came down to disregard it[3]. The truck proceeded, spraying a mixture of diuron and bromocil, mixed with an odor-masking agent called "Lo-scent." No warning had been given to residents that their no-spray signs would not be honored, nor had any notice been given that the roadside would be sprayed.

One-half mile past the spot where Cassens had sprayed my children in 1975, Paul Merrell caught up with the spray truck. He confronted Cassens and his co-worker with a rifle and requested them to delay spraying the other side of the road until he could seek a court injunction to prevent it. Cassens refused. Merrell threatened to use "the minimum force necessary to stop them" should they attempt to return through his farm. Other valley residents had been alerted to the spraying by this time; armed with rifles, they had set up a roadblock around the next turn. Cassens sprayed no further upriver that day. Two weeks later, Merrell was charged in state court with "obstructing governmental administration."[4]

A week after the incident with Cassens, another county spray truck made its way slowly up the Five Rivers Road equipped with an air compressor and spray equipment. At a farm upriver from where Merrell had encountered Cassens, Mrs. Rose Killian heard the spray rig coming. Her children were picking greens for the pigs by the roadside that morning, out of sight of the farmhouse. Fearing they would be sprayed as she knew other children had been in the past, she fired a rifle shot into the air to warn them to come to the house.

Mrs. Killian ran down the driveway to meet the children, only to discover that the truck was not spraying herbicides but spray-painting the center line on the road. Embarrassed, she explained her fears to the crew, and apologized for startling them. ("Oops-Wrong Truck" was the headline on the story in the Lincoln Log, a Newport paper.[5]) Two days later, Mrs. Killian was arrested and charged with "menacing." In October, a jury found her not guilty after hearing testimony about county herbicide incidents that justified her actions. Soon after Mrs. Killian's acquittal, the county began negotiating to drop the charges against Merrell.

Over the years people in many areas have struck directly at the equipment that threatens them with the insidious violence of poisons. Tires of spray vehicles and herbicide tank trucks have been hacked with axes, tools stolen from helicopters, shots fired at them. In 1979, a crowd of angry housewives in Five Rivers abandoned their kitchens when word came via telephone and C.B. networks that a Forest Service tank truck and helicopter crew were approaching the valley. When the Forest Service crew evaded their roadblock by taking a 30-mile detour, the cavalcade of valley women, men, and children scouted the area, located the helicopter landing, and confronted the crew with jeers and threats.

In Minnesota, a local jury had found Harmon Seaver to be justified in firing a rifle at a spray helicopter to defend himself and his family from the chemicals[6]. Ken Riley in Oregon was also acquitted for firing a shot to alert his neighbors that a spray truck was approaching[7]. In the Siuslaw Forest in 1980, an Evergreen Helicopters spray helicopter was "trashed," its instrument panel ripped out and every accessible wire and connection severed. Spray-painted across the body of the helicopter was the message, "Fuck the

Image by Dana Olsen

Helicopter sprays herbicides in Oregon in 1979

Shah." (Evergreen – commonly known as a CIA-connected operation – had recently flown the Shah of Iran to Egypt.)[8]

During the early months of 1981, at least two "Fishline Alliance" flyers appeared at several locations on the west coast, graphically depicting various methods of disabling or blowing up spray vehicles and helicopters[9]. Ron Arnold, a communications consultant and writer specializing in environmental conflicts, reported:

> In the spring of 1980, another feature was added to the anti-pesticide repertoire: terrorism and sabotage. On Sunday, May 4, a Forest Service ground crew was attacked by protesters, beaten, their spray gear destroyed, and their leader forced at knifepoint to sign a statement promising not to spray again in that area...A spray helicopter hangared near Cave Junction, Oregon, had its electrical wires, oil and fuel lines cut. Near Somes Bar, California, a spray crew bus had

its brake lines cut. At Dennysville, Maine, in 1979 an armed mob threatened to shoot down spray helicopters. Someone in Myrtle Point, Oregon, punctured ten 55-gallon drums of Verton 2-D and twenty-eight 5-gallon drums of Tordon-K. Activists in Salyer, California, contaminated 500 gallons of herbicides to prevent their use. Mobs of activists routinely trespass in areas scheduled for aerial spraying and prevent thousands of acres of government lands from being sprayed. In fact, I found a notice in the Wolf Creek Tavern in Wolf Creek, Oregon...that advertises a "Training Weekend in Non-Violent Resistance" where activists are trained to become "occupation forces" to prevent herbicide spraying on government lands. This civil disobedience and sabotage wave has become a routine part of the anti-pesticide movement in the last year.[10]

In 1981, the U. S. Forest Service added an Intelligence Division and Forest Service special agents to its spray program, to monitor local resistance to herbicides and the activities of local "leaders" in the antispray movement[11].

On Saturday, May 30, 1981, the night following Paul Merrell's arraignment, a Western Helicopters spray helicopter contracted by Publishers Paper to spray clearcuts in the Toledo area was completely destroyed by fire. On Monday morning, two women appeared on television and radio news shows, representing the People's Brigade for a Healthy Genetic Future. On camera, they wore scarves and held branches in front of their faces. Following a news report of the burnt helicopter, the two women explained the reasons for the group's actions:

FIRST VOICE: On Saturday, May 30, our Second Annual Spray Helicopter Destruction Derby was completed. A helicopter was burned as a message to the companies who profit from spraying poisons indiscriminately with disrespect for human and animal life and Mother Earth. We oppose violence to all life forces. We sabotage poison-spreading machines as an act of self-defense. We have seen ourselves, our families, and our neighbors suffer ill effects over the years from herbicide contamination, including miscarriages, birth defects, and cancer.

SECOND VOICE: Last fall, we saw international corporations such as Dow, Chevron, and their front organizations distort the election process by spending roughly $100,000 to defeat two local measures intended to safeguard our health. Herbicides and other toxic chemical sprays will eventually be halted when people learn the true cost/benefit ratio. Our present health and genetic future are at stake. We will strike again[12].

Three months before the helicopter incident, Dave Dietz, the pro-pesticide lobbyist who coordinated industry efforts to defeat the Lincoln County initiatives, had told an agricultural chemical conference that public opinion in Oregon was heavily against pesticide use[13]. Strong public opinion is not generally expressed in the burning of helicopters, however. What drives ordinarily peaceful, law-abiding citizens to such drastic action? In the Siuslaw National Forest, perhaps it is simply their own experience. Herbicides have been used experimentally in the area since at least 1948[14]. Intensive management of this highly productive forest has escalated the spraying to a twice-yearly rain of herbicides in nearly every watershed. Over the years, things began to add up.

First, there were the miscarriages, unrelated at first, isolated. After all, how many women are pregnant in a given area at a particular time – the time of the spraying? Over the years, though, they add up. Add the bleeding and the hemorrhaging, the increasing numbers of young women requiring hysterectomies. Add the children sick, the meningitis outbreaks, the pneumonia, the cancers in young people. Add the doctors, confronted again and again in delivery rooms with deformities that most obstetricians rarely see in their lifetimes.

Over the years, they add up, in spite of what some "experts" say. And over the years, people come to know. They read the letters to the editor, they hear the news, they talk with their neighbors, their fellow-workers, with the mothers at preschool, the gas-station attendant, the checkstand clerk, the Bible study group, or their drinking buddies at the local tavern. From the watersheds of coastal river valleys, news of spraying activities and ill effects becomes a topic for gossip in the "grapevine" of truckers, loggers, C.B. networks, housewives, and town residents along the coast of Oregon. While experts juggle laboratory data and administrators ponder their significance, the people of the forest see more and more clearly the link between spraying and illness, hysterectomies, miscarriages, birth defects, cancer, sterility.

"These are your friends and neighbors these things are happening to," says Dee Vawter, who gave birth to an anencephalic child in December 1976. She is a quiet, pretty young woman, working now at a small bookstore in Newport. "We had never heard of these things before. When the doctor told me my baby was going to be born with this defect, I didn't want to believe him. I got angry with him, and wouldn't go back to him again. We went all the way to Portland to have the baby, but my doctor had been right. She was born without any brain. The doctors there wouldn't let us look. They took her away, and we stayed in a motel for nine days, while they kept her alive somehow. We were just there, waiting for her to die – somehow we couldn't just go away.

"We thought we were all alone with that, and then we found out about other people, and about the spraying they do in the forests, in the city watershed. I don't know why these things happen. Does anyone? I don't want to believe the sprays could cause these things. But I didn't want to believe my doctor, either, and not believing didn't change what happened."[15]

The people come to know. Politics, meetings, courtroom games, experts and statistics, organizations – all are meaningless when the helicopters come over the ridge. When people know enough, there is no further use for talk.

"Suppose someone points a gun at your child," says a young mother, loading diapers into a machine at the laundromat. "You yell at him to stop, and he says don't worry, it's just a little bullet, a safe kind of bullet. Good grief, how can there be a safe bullet, you say, and he says he has scientific evidence it's safe. So show me your evidence, you say, and he says no, it's secret information. Are you going to stand there and let him shoot your kid?"

Picture it, then. It could have been a scene from a movie.

In a clearing on a ridge above Toledo, the helicopter has been sitting for a week, grounded by bad weather. The clearing is secluded, but even from a distance the glint of metal and Plexiglas, the ghost of rotor blades behind the trees are visible to a keen or searching eye. No guards attend the machine, and for a few days no one notices it. Then some teenagers spot it, perhaps, joy-riding on the logging roads after school. Perhaps a fisherman notices it, or a young girl on horseback, a family out gathering firewood, a laid-off mill worker scouting for cedar.

Within hours, the helicopter's location is known up and down the coast. In the bars, cafes, laundromats, wherever people gather, it is discussed, its location noted, oblique references made to its fate. "You mean there ain't even a guard on it?" says an old fiddler at a local tavern. "Why, they hadn't ought to leave it like that, it might just give a fellow ideas." He improvises new lyrics for his next tune, "The Baltimore Fire," and everyone laughs.

Rain and fog continue to ground the chopper. As the weekend approaches, various hikers, motorcycle riders, teenagers in four-wheel-drive pickups, bird watchers, wood-cutters, and others appear casually in the dismal weather, study the helicopter, and leave. One afternoon a Winnebago camper parks discreetly off the highway, and an elderly couple hike through the rain to the landing pad. They stare at the helicopter and study the surrounding terrain, then slog through the mud and ruts of a logging road back to their camper.

On Saturday afternoon, the weather clears. Over the C.B., at the taverns and cafes, the change is mentioned, the possibility of spraying the following day. Spraying is weather dependent, and people know from past years that Sunday is as good a day as any other to a helicopter. Word comes from a trucker, perhaps, or a motel worker or even an "insider" at Publishers Paper that the word is go if the weather holds.

There is no organization, no plan. It just happens. In the clearing, the helicopter sits, the light of the waning quarter moon glimmering on its plastic bubble. Inside the cockpit is an old woman. She wears a large black cape and a fur hat, pulled snugly down over her ears. She is twisting newspapers into logs, arranging them on the floor of the cockpit.

The forest silence is broken by the sound of a vehicle. Headlights loom brokenly through the trees. Some distance from the landing pad, the sound of the car motor stops. The old woman crouches in the cockpit, waiting, listening. The forest is still again. Twenty minutes later, four figures in dark clothing appear at the edge of the clearing. Their faces are blacked with mud or soot. Two of them are teenage boys, one holding the hand of his girlfriend, the other carrying a bundle of dry sticks and twigs. With them is a woman, wearing a jogging suit with a local high school emblem on it. She carries a can of gasoline. They stand hesitantly in the shadows. Twice the boys start forward, but the woman gestures them back.

Letting go his girlfriend's hand, one of the boys picks up a slab of fir bark and heaves it across the clearing. It misses the helicopter and strikes the steel casing of some radio equipment with a hollow thud, as startling as rifle fire in the silent forest. The radio antenna rocks and vibrates. No lights or voices respond. The boys look at the woman, who glances around once more and nods. They break cover and run toward the helicopter.

The cockpit door opens suddenly, and a flashlight blinds them, scanning their startled faces, the pile of sticks, the can of gasoline. The old woman clambers heavily from the cockpit, hindered by her cloak. She is laughing. Muffling the flashlight with her hand, she leads the four on a guided tour of the helicopter, displaying her handiwork in the cockpit. The boys are laughing, too. One takes the gas can from his mother, and fumbles with the cap.

The sound of another vehicle approaching stops him. He sets down the can and grabs his girlfriend's hand. The old woman slams the cockpit door shut. All five run into the brush at the edge of the clearing.

A pickup whines in low gear and stops, its headlights flooding the clearing. In front of the helicopter, the small red gas can casts a long shadow. Three men hop out of the truck, each with a beer bottle in his hand. Two of them wear heavy caulked loggers' boots. Their pants are held loosely with suspenders, and the cuffs are frayed. The third man is shorter than the others. He wears a leather cap and walks with a limp. The pockets of his field jacket bulge with unopened bottles of beer. He taps the gas can with his foot, noting its fullness, and picks it up.

The old woman and her companions plunge out of the brush.

She takes the gas can from the man in the field jacket, gesturing defiantly at the helicopter. She wags a finger at the men and points at the pickup. One of the loggers trots obediently to the truck, douses the headlights, and drives it with only the parking lights on to a more secluded spot. He returns to the helicopter, where the others are sitting, the man in the field jacket distributing beers from his pockets.

A flashlight flickers suddenly in the trees beyond the clearing. As if on cue, the group fades into the brush opposite, herded noiselessly by the man in the field jacket. He limps after them, picking up an empty beer bottle and the gas can on the way. The flashlight approaches and goes out. A young couple stands in the moonlight, peering at the helicopter. Each wears a backpack and sleeping bag, and the young man carries a can of charcoal starter...

The crowd grows. A trucker arrives with wrenches, detonation cord, and a quart of beer. Teenagers, motorcyclists, hikers appear, and three women from a coastal tavern arrive with more beer, another can of gas. A middle-aged couple wearing down vests and Hallowe'en masks bring two thermoses of coffee and a can of Coleman fuel.

From the west, the quarter moon casts long shadows among the whispering crowd. Cigarettes glow, bottles clink, accompanied by muffled laughter, hushed conversations. The old lady and the man in the field jacket move among the shadowy figures, urging them not to leave cigarette butts, matches, bottle caps. A shooting star streaks overhead, leaving a glowing trail. From somewhere in the darkening sky drifts the faint gabbling of a late flock of geese, heading north.

The party breaks up, shivering. Cigarettes are snuffed and pocketed, bottles collected. Gasoline and other flammable offerings are poured on tinder piles in and below the helicopter. The cockpit reeks of fumes, the smell filling the clearing. Carrying thermoses, bottles, and empty fuel containers, the party retreats to the trees beyond the clearing to discuss how to light the pyre. Alone in the moonlight, the man in the field jacket limps back and forth, in and out of the helicopter's shadow, inspecting the clearing with a flashlight and scouring the ground with a heavily branched alder sapling.

One of the loggers leads the old lady back into the clearing, whispering instructions. The man in the field jacket, his inspection completed, joins them. The logger strikes a flare. He bows to the old lady and presents it to her. She holds it gingerly and throws. Her cape impedes her and the flare falls short, casting a festive red glow on the three figures and the helicopter. The old lady laughs.

The logger strikes another flare and hands it to her. She flings her cape back, winds up like a major league pitcher, and hurls. The arc of the flare ends in a burst of flame from the cockpit doorway of the helicopter. The entire machine ignites simultaneously. The blinding flash and deep, breathless thump of the explosion pierce the forest. The blast knocks the old lady's fur hat off and flings her to the ground. The logger yanks her toward the forest. His companion, still dragging the heavy sapling, picks up the old lady's hat on the run. Limping, he follows them toward the trees, brightly lit now by flames.

Behind them, the helicopter erupts. Plastic and frail metal alloys swell and buckle in a searing halo of white and orange. An explosion topples the rotor from its mount. Among the trees, the crowd watches. The tail of the helicopter, a skeleton in the flames, curls up and over like a scorpion. After a single shout of triumph, the crowd is silent, awed by the immensity of the fireball from the crumpled bit of machinery.

Like a Fourth of July crowd after the fireworks are over, the party disperses, hastening through the forest in different directions. Their footsteps fade, the sound of the last motor dwindles in the direction of the highway. In the clearing, the frame of the helicopter slumps awkwardly, still glowing amidst small pools of flame. Trees beyond the perimeter rustle in the small wind from the fire. The flames subside to glowing embers, sputter briefly to life once or twice, and the forest is still once more.

* * *

Destruction of an aircraft is a federal crime. The investigation of the helicopter burning involved the FBI as well as local county police, who delegated a special investigator to the job. Publishers Paper offered a $5,000 reward for information leading to the arrest of the persons responsible[16]. A spokesman for the company, Bill Lesh, told reporters, "This is a destruction of private property. It's an assault. You don't see us going around burning out homes. It's not the way we go about things in a democracy."[17]

The investigators apparently found few leads to pursue at the site of the burning and concentrated their efforts on attempts to identify the two women who had made the "People's Brigade" statement on television[18]. Even if they could identify the two women, however, there was no assurance that the two were actually involved in the incident.

On the street, in public places from Coos Bay to Tillamook, the investigators might have found an abundance of suspects. As soon as the incident was reported in the news, it became almost the fashion to boast of having been involved.

"If everyone who says they were there really was, it must've been some party," one Newport fisherman commented.

As time went on, tales were embroidered, characters added, conflicting details supplied. Response to such tales was skeptical, but generally sympathetic. Clearly, the destruction of the spray helicopter – however it was accomplished – was an acceptable act to many people.

"My only regret is, I wasn't in on it myself," said a man in a suit and tie during a theater intermission in Newport. "I'd drop everything and come all the way from Salem to get in on the next one."

"Where's this guy coming from, anyway?" says a woman at the check-stand at Safeway, reading the statement of the Publishers Paper spokesman. "So they don't burn our houses, they just poison our babies – that's his idea of democracy!"

"They've had this power over our lives." another woman says, "The power to spray, to bomb, to build their LNG [liquid natural gas] tanks and pipelines and nuclear plants, and what have we been able to do to stop them? It's only a small thing – one little helicopter – but it's a little chip of power we've been able to take back."

Outside a chainsaw repair shop a young logger shares a thermos of coffee with a friend, waiting for his saw. His sandy hair is cut short, and the bumper sticker on his pickup reads, "Sierra Club Kiss My Ax."

"Hey, I'm no environmentalist." he says. "Don't get me wrong. But I'm no dummy, either. They sprayed our watershed almost every damn year. My first

Image by Rio Davidson

Old growth forest in Oregon's Valley of the Giants

kid was born with a clubfoot and had to have three eye operations. And that's the only kid we'll ever have, because my wife couldn't get pregnant after that. She had so many problems these last few years she had to have an operation. They come around after they spray – afterward, for Chrissake, without ever telling us ahead of time what they're spraying, or when – they come around and test our drinking water, but when I ask what they found in it they won't tell me. There's a whole raft of shit they're not telling us.

"Nope, they don't fool me, trotting out their experts. I know which side their bread is buttered on. So help me God, if they come around my water supply again without asking, I'll shoot 'em out of the sky. They're crying in their beer over one burnt up helicopter." He laughs. "I could tell 'em a thing or two about that. But if they can keep their secrets, I guess I can keep mine."

"Right on," his friend says. "I sure as hell wouldn't go advertising what I know on TV. If they come near my place, they'll never know what hit 'em and they'll never know who did it."

At a high-school football game, a woman watches her son emerge from a heap of sprawling, muddy bodies. She sits down as the cheers subside and continues her conversation with a friend.

"That's what the spraying is, too – another macho compulsion. Men have this compulsion to control the earth – the air, the water, the land, anything that moves or grows. They've lost touch with life. Life isn't the important thing to them, what they want is power over life. That usually means the pow-

Image by Peter von Stackelberg
Mist shrouds the forest along Highway 34 through Oregon's Coastal Range

er to destroy. I don't think women would have invented all these things – bombs, chemical weapons. They're closer, more in tune with the earth, with little things, with elements of life that can't be controlled – except by death, the ultimate control..."

"They're destroying the world," the woman with her says, "And they make the laws that protect them while they do it. And we're responsible too, for sitting by while they do it. We teach our kids to be law-abiding, but we have to teach them that the law isn't necessarily right – or that something that's illegal isn't necessarily wrong." The women laugh. "We've taught them one thing, anyway – when push comes to shove, we've got some power, too."

"I don't want my kids to grow up and be part of a system that solves its problems by killing," the first woman says. "I don't believe in that. And I don't believe it's wrong to destroy machinery that kills – that's designed to kill – trees, children, animals, the salmon coming in from the sea. It means breaking their laws. I accept that, because I don't accept a law that values a machine or profit over life and unborn children."

"Laws can change," her friend says. "Maybe our grandchildren won't have to go around blowing up helicopters."

Informed Discretion

*If the Bill of Rights contains no guarantee that a citizen shall
be secure against lethal poisons distributed whether by private
individuals or by public officials, it is surely only because our
forefathers, despite their considerable wisdom and foresight,
could conceive of no such problem.*

—Rachel Carson

The story of the fight against herbicides is the story of some very ordinary
individuals – housewives, veterans, farmers, forest workers, clergy, parents –
who saw something very wrong and set out to correct it. In the process, they
discovered some extraordinary capabilities within themselves. Their story is
above all a chronicle of what went wrong after Silent Spring.

The powerful individual and social consciousness awakened by Rachel
Carson found expression in the National Environmental Policy Act of 1969
(NEPA), in amendments to the pesticide law (FIFRA) ordering rigorous
safety testing of chemicals, and in the creation of the Environmental Protection Agency. These monuments to Rachel Carson were to be the public's
guarantee that her silent spring would not occur.

The response of the chemical industry to the new laws created to ensure
the safety of pesticide products was simply to falsify or suppress laboratory
evidence of their poisonous effects.

At the same time, industry engaged in massive public relations campaigns
and industry-sponsored university research programs designed to build public trust and dependence on chemicals – in other words, to reprogram the
consciousness Carson had inspired. The EPA was handicapped from its birth

by poor organization, a backlog of outdated and inadequate test data, lack of funds, and the transfer of personnel and chemically oriented philosophy from the United States Department of Agriculture[1]. The fledgling agency was easy prey for powerful industry influence.

The laws intended to protect human health and the environment from toxic chemicals define clearly the rights of manufacturers of the poisons, but nowhere are the rights of the public firmly established[2]. NEPA states not that each person shall, but that "each person should enjoy a healthful environment."[3] Thus Congress determined a healthful environment to be not a right, but merely a wistful hope. The courts have interpreted NEPA to mean that the public can sue federal agencies only to enforce proper procedure; a foolish agency decision, if procedurally correct, cannot successfully be challenged[4]. The public has no right at all to sue for enforcement of FIFRA[5].

Not only has the public's right to sue been denied and limited, but also its right to know. EPA has consistently prevented public scrutiny of health and safety information on pesticides, in defiance of U. S. Supreme Court decisions which affirm the public's right to know[6].

Operating behind a nearly impenetrable cloak of secrecy, pesticide manufacturers and the federal agencies that both regulate and use their products have effectively subverted the laws and institutions created to ensure that Rachel Carson's grim specter should not become a reality. Congress and the American people have failed to exercise their legal rights to full disclosure of information and active participation in decisions on pesticides. Without the ballast of public pressure, bureaucracies are vulnerable to the overwhelming pressure of corporate interests. The right to know and the right to participate are the public's only safeguard against governmental improprieties.

The public is consistently discouraged from examining or participating in governmental activities, however. No agency has been more diligent in this regard than EPA. Entrusted with the noble task of protecting the environment and human health, EPA became a bureaucracy of concealment, a house of cards built on the premise that what people don't know won't hurt them. There are possibly kinder explanations for EPA's behavior, but none explain the agency's actions upon discovering widespread fraud in registration testing. Secrecy became a necessary part of the regulatory process, to cover up the fraud and negligence underlying pesticide registrations.

Watergate should have taught us the dangers of secrecy in government. Public awareness is the lifeblood of democracy, as James Madison recognized in 1822:

> A popular Government, without popular information, or the means of acquiring it, is but a Prologue to a Farce or a Tragedy; or, perhaps both. Knowledge will forever govern ignorance: And a people who mean to be their own Governors, must arm themselves with the power which knowledge gives.[7]

Forest residents and Agent Orange veterans are not the only victims of regulatory corruption and secrecy; because of their relative isolation from other forms of pollution, their sufferings are only more visible and identifiable. Other victims may never suspect the causes of their health problems. Without public consent and largely without public knowledge, the regulatory process has permitted well over a billion pounds of poison to be applied annually within the United States[8]. It is impossible to live anywhere in North America without being exposed to many of these substances[9].

The same herbicides that are sprayed on our forests are used on school playgrounds, golf courses, parks, the lawns of public buildings, roadsides, and in lawn and garden products sold to consumers. Fruit, produce, meat, milk, and other foods sold in supermarkets may contain "allowable" levels – determined in secret – of a great range of insecticides, fumigants, herbicides, fungicides, and other poisons, many of them fraudulently and inadequately tested. Buildings, commercial aircraft[10], and foodstuffs may be routinely fumigated, and many building materials are treated with highly toxic preservatives, which vaporize into the air of home and workplace. Treated fabrics, cosmetics, deodorants, disinfectants, food additives, food containers, flea collars, and other common household items all add to the toxic burden individuals begin to accumulate before birth.

Residues from manufacture of these poisons accumulate in secret toxic dumps, where some break down into still more deadly compounds. Industrial, agricultural, and radioactive wastes seep into water, soil and air, carried far and wide by winds, rivers, and ocean currents. Political judgments about the hazards of these poisons may change, but toxicity does not. Government regulations have failed to protect humans from exposure. Pesticide residues and their breakdown products are found routinely in blood and urine samples of both urban and rural people nationwide. In every sample tested of human semen and sperm, pesticide residues have been identified which are capable of causing mutations and cancer[11]. The same substances turn up in mothers' milk[12].

The little that is known about some of these pollutants would not persuade a reasonable person to eat, drink, or breathe them willingly. In addition to birth defects, genetic damage, cancer, and other devastating effects, many of these poisons cause behavioral effects, such as anxiety, tension, depression, learning disorders, hyperactivity, insomnia, irritability, and other problems, most commonly attributed to "psychological" causes[13]. While the total physiological effects of such poisons on an entire society may be incalculable, the possibility of whole populations or nations coming unhinged psychologically is truly apocalyptic.

Could it be that chemically induced imbalances in brain chemistry distort people's reason to such an extent that they are unable to defend themselves against the poisons which cause such effects? Certainly a society which inflicts mind altering, deadly poisons upon itself in the name of economic pest control is not behaving rationally.

The enormity of the knowledge of countless poisons invading our lives, leaving unknown toxic legacies for future generations, is too appalling to contemplate. So long as people are kept ignorant, so long as they can be convinced that only government and "experts" can solve the toxic nightmare, poisons blessed by regulatory approval will continue to accumulate. Officially enforced public ignorance has rendered people powerless and apathetic. By the time toxic effects become too obvious to ignore, it may be too late.

For some of us, it is already too late. In the sprayed forests of the Pacific Northwest, at least, birth defects, cancer, sterility, and miscarriages have deprived people of their unalienable rights to pursue a healthy life, to bear children, and to hope. Government and industry, knowing that their laboratory tests are inadequate and their registrations a sham, refuse to accept the "circumstantial" evidence of toxic effects in forest populations, Agent Orange veterans, the people of Vietnam, of Globe, Arizona, of Ashford, Washington, and everywhere else these chemicals are sprayed.

Sooner or later, however, circumstantial evidence becomes too strong to deny. In the days when farmers sold milk directly to consumers, a purchaser had no way of telling whether or not the milk had been "watered," but – as Thoreau suggested – finding a trout in the milk was ample reason to buy milk elsewhere. The human subjects of the forest chemical experiment are, like Thoreau's trout in the milk, a message to the nation that regulation of poisons has failed, that it is time to look elsewhere for a solution.

The answer lies not in regulation and enforcement, but in returning the control of poisons to its democratic source: the people. Rachel Carson suggested the remedy: "If the Bill of Rights contains no guarantee that a citizen shall be secure against lethal poisons..."[14], then it is time to add that guarantee to the Constitution. New technologies unforeseen by our nation's founders require an addition to the Bill of Rights for those who bear the risks.

An informed-consent amendment could provide that no law nor court shall limit or impair the people's right to obtain on a timely basis full disclosure of information about matters which may affect them; nor except to protect from a clearly greater harm shall any person be unnecessarily deprived, in time of war or peace, of the right to grant or withhold informed consent before being exposed to a potentially harmful substance or energy form. Economic necessity should not be a determining factor; however, language can surely be drafted that adequately and fairly addresses "necessary" pollutants such as automobile exhaust. For example, Congress could be given authority to establish abatement requirements and classify individual pollutants as "essential" for set periods of time, thus assuring that pollutants are actually reviewed. Under such a scheme, responsibility for developing abatement plans and exemptions should not be turned over to a regulatory agency, but should be retained by Congress so that such issues will be resolved through the democratic process.

Under present law, a citizen threatened with exposure to a toxic chemical is required to prove that irreparable harm will occur while the information necessary for such proof is kept secret or is simply not available because the research has never been done.

A responsible physician would not consider prescribing a drug without first clearly explaining to the patient any possible side effects of the medication and warning of any toxic reactions if it is taken with alcohol or other drugs. A doctor who fails to make the facts clear to the patient, or a drug manufacturer who fails to instruct both physician and patient, can be held accountable for any illness or injury that ensue. With full knowledge of a drug's effects, the patient may choose whether or not to accept the risk.

Drugs and pesticides are biologically active substances which differ only in name and in the matter of consent. Informed consent is already a recognized legal principle with regard to medical patients, experimental human subjects, tort case law (the doctrine of strict liability), and the labeling of voluntarily used products, such as drugs and cigarettes. Cigarette smokers, however, currently have no right to know the identity or possible effects of some 300 additives in American cigarettes, or the unknown pesticides used on the tobacco. The informed consent of parents is required for school projects involving their children. Extending the principle of informed consent to situations that now result in involuntary exposure (e.g., pesticide residues on foods, contamination of soil, air, or water, radiation hazards) is particularly appropriate. Those who bear the risks of such exposure do not enjoy a proportionate share of the benefits.

The campaign for this reasonable constitutional right could provide a positive focus for the many citizen groups now exhausting themselves in separate battles for the same thing. The resources of veterans' groups, environmental groups, clergy and religious organizations, physicians and medical associations, attorneys, health-related societies, labor unions, antinuclear activists, and other diverse groups sharing common concerns could be engaged in massive public-education programs centering on the issue of informed consent.

Regulation of pollution has escalated the problems of pollution by fragmenting the issue. Innumerable separate laws and regulations for each separate element of pollution have managed to divide and conquer the powerful social consciousness Rachel Carson awakened. The public must fight countless battles piecemeal, on a case-by-case basis. During the battle over a single chemical such as 2,4,5-T, a host of new poisons appear on the market. Vital decisions about safety and environmental impacts of poisons are subject to political and economic whims, and any progress toward a sane pesticide policy can be set back years by a change in political administration. The principle of informed consent transcends all these difficulties.

For both citizen groups and individuals, informed consent demystifies the gap between society and decision-makers who presume to act in the public interest. Its very simplicity gives the public access to complex subjects which have become the exclusive domain of legal, scientific, and bureaucratic "experts." These experts know no more about informed consent than anyone else. A welder, poet, housewife, or fisherman can successfully confront any expert by addressing the basic issue: "Do I or do I not have the right to know the effects of the stuff you intend to expose me to? Do you or do you not have to get my permission before exposing me to it?" These questions could become the focus for school, city, county, and state voter initiatives.

A constitutional amendment would not be an instant solution to the problems of pesticides and other economic poisons. Ratification is a slow and tedious process. But the Equal Rights Amendment demonstrates the effectiveness of that process. Vehemently opposed factions have focused attention on women's rights. The E.R.A. is still far from being ratified, but the campaign for it has so successfully educated the public that, whether it passes or not, many women have come to insist on their rights, and society is beginning to recognize them.

Similarly, an informed-consent amendment may take years to be ratified, but in the course of its campaign, people would come to exercise their right to informed consent and society would recognize its validity long before it became signed into law. As consumers exercised their right to know about poisonous residues in their food, buying habits would change, and marketing changes would follow. Such changes have already occurred in the marketing of brand-name foods containing no preservatives. In response to consumer awareness and buying habits, major food processors now promote their products on this basis. Pesticide residues in food could be as readily affected by consumer buying habits. Higher demand would provide incentives for farmers to grow food without pesticides, which would in turn affect the direction of agricultural research.

The issue of informed consent would give the public the right to weigh the risks of poisons against their benefits. The vast resources now wasted in futile regulation will never achieve the effectiveness of individualized, personal control over poisons. Instead of politically vulnerable government agencies making such decisions in secret, based on secret information, the public itself could decide whether to accept the risk of cancer, birth defects, genetic damage, miscarriages, neurological and other disorders in return for redder apples, dandelion free lawns, cheaper lemons, or a higher allowable timber harvest.

To those who bear the risk, exposure to poisons is a matter of life and death. Decisions about poisons which disrupt the basic mechanisms of life and the integrity of future generations raise profound moral questions. Society cannot afford to entrust such decisions to corporate or political entities that exempt themselves from the restraints of morality and ethics.

Informed consent is a positive, effective remedy for the catastrophic failure of regulation to protect the public from lethal poisons. Diverse political, social, religious, and public interest groups, by agreeing on a single issue, could not only enlighten society about an awesome danger, but most importantly, would give people something they can do about it. A doomsday approach to the problem offers people no more hope than knowledge that a meteorite is about to strike. Informed consent gives people not only the right to know, but also the responsibility to act, and above all, the means to act.

The regulatory process would then become what it should have been to begin with: a public-information service which could itself be held accountable for failure. Scientists who – through incompetence, negligence, or greed—produced grossly faulty research on poisons would also be held accountable. The public's right to clear, complete information would be unequivocal. All safety and efficacy data, with no exceptions, would be made available to the public before a chemical could be registered. Its label would clearly warn of all possible effects and would advise the users that they will be held strictly liable for exposing humans without their informed consent.

Industry and government would understandably find informed consent distasteful. They would proclaim it impractical and expensive, and inimical to public welfare – claims which have little more credibility than unfounded claims of pesticide safety. In opposing informed consent, industry and government would have to admit that toxic exposure does occur and that people have no right either to know about it or to stop it. They would have to convince the public that ignorance and helplessness are preferable to knowledge and control.

A society insisting on its right to informed consent would quickly redirect pest control into healthier avenues of inquiry. All the ingenuity and resources now devoted to developing chemical weapons against pests could be directed toward nonpoisonous methods of integrated pest management. Pest control would adopt nature's system of checks and balances, rather than disrupt it. The manufacture and use of poisons would be subject to the social system of checks and balances essential to a true democracy.

The idea of effective agriculture and forestry without pesticides is more than a pipe dream. The Canadian government, recognizing the folly of its reliance on U.S. testing and regulation of pesticides, is already taking steps to wean Canadian agriculture and forestry from their addiction to poisons. The Canadian Environmental Advisory Council has formulated a political strategy for reducing – and ultimately eliminating – its nation's reliance on chemicals to control pests. The Canadian council's excellent report examined in detail the failure of pesticide regulation and questioned "The appropriateness of using toxic chemicals for the suppression of pests in the first place."[15]

In 1981, the Canadian council completed a comprehensive plan to shift the emphasis from exclusive reliance on chemicals to the "more sophisticated, multi-technique approach" of integrated pest management. The report outlines the political and economic policies which "would permit Canada, with its substantial body of expertise, to minimize the use of chemicals in favour of control techniques designed in Canada for Canadian conditions."

Chemical pesticides, although they control effectively many pests, impose an environmental and human health burden. Because that burden cannot be adequately assessed, the present risk/benefit judgments made in order to register a pesticide are invalid. Imposition of unknown risks could be avoided or minimized by implementing alternative control strategies. A regulatory system focused on chemical pesticide regulation does not encourage such implementation. The time has come to shift emphasis to facilitating total control strategies.[16]

This admirable Canadian proposal, however, fails to offer relief to the individual threatened by a spray nozzle, and assumes that government bureaucracies can successfully withstand the awesome power of the chemical industry. Only a vigilant public, armed with the principle of informed consent, can safeguard those bureaucracies against the same pressures that have corrupted pesticide regulation. So long as one person presumes to make decisions for others, the pesticide issue will not be resolved.

As Rachel Carson noted, the founders of our government could not have foreseen the problems posed by "economic poisons" which benefit a few to the hazard of many[17]. Their wisdom, however, foresaw a remedy. "I know no safe depository of the ultimate powers of society but the people themselves," wrote Thomas Jefferson, "And if we think them not enlightened enough to exercise their control with a wholesome discretion, the remedy is not to take it from them, but to inform their discretion."[18]

A government which truly derives its just powers from the consent of the governed cannot deny its people the opportunity to "inform their discretion" and to exercise their control over potentially lethal poisons. The people have an equal duty to exercise their rights. More is at stake than the irretrievable erosion of democracy. Against the deadly proliferation of economic poisons, only an informed, caring public can defend the greatest gift of all – life itself.

Appendix A

The information in Appendix A is reproduced photographically, reduced in size, from U.S. Air Force documents distributed to the U.S. Veterans Administration Medical Advisory Committee on Agent Orange at Saint Louis, Missouri, on February 5, 1980.

COMPONENTS OF SELECTED HUMAN SYMPTOM/SIGNS FOLLOWING EXPOSURE TO PHENOXY HERBICIDES AND/OR TCDD

NEURO-PSYCHIATRIC ABNORMALITIES

ASTHENIA
ANXIETY
DEPRESSION
FATIGUE
APATHY
LOSS OF DRIVE
LIBIDO
IMPOTENCY
SLEEPLESSNESS
EMOTIONAL INSTABILITY
ANOREXIA
DIZZINESS
↓ LEARNING ABILITY

PERIPHERAL NEUROPATHY
HYPOREFLEXIA
WEAKNESS
PARESTHESIAS
EXTREMITY NUMBNESS
MYALGIA
GAIT DISTURBANCE
"MILD" PARESIS

DERMATOLOGIC DISEASE

CHLORACNE
PORPHYRIA CUTANEA TARDA
HYPERPIGMENTATION
HIRSUTISM (BODY)
ALOPECIA OF THE SCALP

COMPONENTS OF SELECTED HUMAN SYMPTOM/SIGNS FOLLOWING EXPOSURE TO PHENOXY HERBICIDES AND/OR TCDD (CONT'D)

OTHER DISORDERS

HEPATIC DYSFUNCTION
↑ CHOLESTEROL
↑ SGOT , SGPT, LDH

RENAL DYSFUNCTION
PROTEINURIA
↓ OUTPUT
TUBULAR DEGENERATION
GLOMERULAR DEGENERATION
RENAL GLUCOSURIA

GI DISTURBANCE
NAUSEA
VOMITING
DIARRHEA
GASTRITIS
ABD PAIN
FLATULENCE

CARDIAC DISTURBANCE
BRADYCARDIA
TACHYCARDIA
ATRIAL FIBRILLATION

Appendix B

Press Release

Statement by Dr. Samuel S. Epstein on Behalf of the Interfaith Center on Corporate Responsibility: "2,4,5-T and the Dow Track Record"

My name is Samuel S. Epstein. I am professor of occupational and environmental medicine at the School of Public Health, University of Illinois, Chicago. I am an M.D., human pathologist, and toxicologist with particular interest in the toxic effects of chemicals and environmental pollutants (particularly pesticides), areas in which I have some 220 scientific publications besides 5 books. I have served as consultant to the Senate Committee on Public Works and other congressional committees, as consultant to various federal agencies, as member of various advisory committees to the Environmental Protection Agency for about 9 years, and as expert witness to various federal agencies.

In 1969, I was appointed by HEW Secretary Finch to serve on several panels of his "Commission on Pesticides and Their Relationship to Environmental Health," and chaired the panels on birth defects (Teratogenicity) and genetic effects (Mutagenicity) of pesticides.

A. Hazards of 2,4,5-T and Silvex

In 1969, the HEW Teratogenicity Panel unanimously concluded, based on evidence then available, that 2,4,5-T and related derivatives (esters) induce birth defects (teratogenic) in experimental animals and that their use "should be immediately restricted to prevent risk of human exposure."

In the succeeding decade, the scientific basis of these recommendations has been overwhelmingly confirmed and extended by a wide range of additional findings, including the following:

Over a dozen studies have confirmed the teratogenicity of 2,4,5-T and silvex in experimental animals (birth defects being produced in most of these studies at the lowest dose tested).

Dioxin, a stable and persistent contaminant, is present in commercial preparations of 2,4,5-T and silvex at concentrations of about 100 parts per billion. Additionally, relatively high levels of dioxin are produced by combustion of 2,4,5-T and agricultural products or other materials contaminated with such herbicides.

Dioxin produces birth defects and reproductive abnormalities at the part per trillion level. Dioxin is the most potent known teratogen, being over one million times more active (in various species) than thalidomide.

Dioxin is the most potent known cancer-producing agent (carcinogen), being active at the part per trillion level.

Preliminary information on a group of Monsanto workers involved in a dioxin accident in 1949 has found an excess of lymph cancers (lymphomas) among those dying since 1969.

A series of recent studies in Sweden have reported a marked excess of cancers (soft-tissue sarcomas) in railroad and other workers exposed to phenoxy herbicides, particularly 2,4,5-T (Hardell & Sandstrom, 1979).

There is increasing evidence of large-scale public exposure to 2,4,5-T, including the identification of dioxin residues in beef fat.

Based on such evidence, and supported by a wide range of the expert independent scientific community, EPA has recently suspended most agricultural uses of 2,4,5-T.

B. The Position of Dow on 2,4,5,-T and Silvex

As overwhelming scientific evidence on the hazards of 2,4,5-T and dioxin has accumulated in the last decade, Dow has attempted by every possible means to challenge and discredit them, and to insist on unattainable degrees of scientific precision.

The excessive level of proof Dow demands for such demonstration of hazard seems inconsistent with the low level of proof Dow is willing to accept for the benefits of its products, such as the alleged economic benefits of the use of 2,4,5-T (see J. M. Newton, 1979). Dow has repeatedly challenged the relevance of experimental animal data to humans, although Dow continues to insist on the acceptance of negative toxicological data on its own products as evidence of their safety. Dow has not only shifted the burden of proof to the public and government, but has also failed to mount appropriate epidemio-logical studies in exposed populations. While sharply criticizing government epidemiological studies, Dow has still failed to publish the results of its own reproductive studies on the wives of some 300 dioxin-exposed workers; these studies were initiated in July, 1976, and concluded by July, 1978.

C. The Position of Dow on Safety Regulation in General

Dow has consistently adopted intransigent positions with relation to safe-ty regulations, such as Toxic Substances Legislation, whose thrust is largely to create the requirement for notification and testing of toxic chemicals prior to their introduction to commerce. Dow has also supported the position of the Manufacturing Chemists Association that workers do not have the right to know the nature and identity of chemicals to which they are exposed in the workplace. Furthermore, Dow has strongly opposed recent proposals of the Department of Labor to improve the currently minimal regulation of carcin-ogens in the workplace.

In this connection, it would seem appropriate for Dow shareholders to enquire about how much of their investments is being spent on support of defensive industry propaganda and public relations, and on support of such trade associations as the American Industrial Health Council, whose activi-ties are largely directed to weakening federal regulation of toxic and carcino-genic products.

D. Dow's Track Record on Scientific Data

Dow's positions on 2,4,5-T and safety regulation are consistent with its track record on the handling of scientific data on the hazards of its products.

In spite of the extensive toxicological studies conducted on its various products, Dow has rarely used such data to protect consumers and workers, but rather to protect its own commercial interests. The record with relation to products such as benzene, epichlorohydrin, ethylene imine, and DBCP (a soil fumigant), besides phenoxy herbicides, establishes that questions on their hazards were first raised and posed by outside scientists or exposed workers, rather than by Dow. For instance, in the 1977 Dow Stockholder's Quarterly Report, it is admitted that "Dow marketed DBCP in 1957 after

6 years of testing and research." Yet, the first indication of its hazards to exposed workers was delayed until 1977, when the workers themselves found that they had become sterile following exposure. It should also be noted that Dow has since withdrawn DBCP from the market, but has replaced it with the more carcinogenic EDB (ethylene dibromide).

The record also establishes the following: Dow has withheld or suppressed its own data on the genetic hazards of benzene and epichlorohydrin and on the reproductive hazards of dioxin; Dow has destroyed its own data on occupational carcinogens such as ethylenimine; and Dow has failed to warn workers on the hazards of sterility of such products as DBCP.

It is also of interest to note the recent resignation, dismissal or premature retirement of senior Dow scientists involved in studies on genetic and reproductive hazards to exposed workers; these studies commenced in about 1968 and were terminated by management in 1977. The specific circumstances of this exodus are indicative of a conflict between the scientists' proper desire to publish their findings on the occupational hazards of Dow's products, and the company's unwillingness to permit such publication.

E. Recommendations

The initiative of the religious investors to require Dow to establish a review committee is consistent with an overwhelming body of scientific data which demonstrate the existence of reproductive and carcinogenic hazards of 2,4,5-T or which pose still unresolved questions. The establishment of such a review committee is not only in the best interest of workers and consumers exposed to or handling Dow products, but is also in the best long-term economic interests of the company and its shareholders.

SAMUEL S. EPSTEIN, M.D.

Appendix C

Dr. M. S. Legator's Address to Dow's 1981 Annual Meeting

My name is Marvin S. Legator. I am speaking on behalf of the Catholic Diocese of Albany, New York, the beneficiary owner of 600 shares of Dow Chemical stock.

I am pleased to appear here today as a private citizen and an advocate for the resolution to establish a review committee to examine and evaluate the potential health consequences of 2,4,5-T, silvex, and their derivatives and to place a moratorium on their production until a report is issued.

During my professional career I have been associated with Dow Chemical as a consultant for seven years. Many of my close associates are either past or present professionals employed by Dow Chemical Company. I have also been associated with a variety of industrial concerns, and I am fully cognizant of the contribution of chemicals to the high standard of living we enjoy in this country.

Given my past affiliations with Dow Chemical Company I was distressed to learn that on April 22 of this year a national environmental organization cited Dow Chemical Company as one of the "filthy five companies" in the United States. This dubious honor was bestowed upon Dow for their unmistakable pattern of repeated serious violations of environmental law and heavy contributions by their political action committee to congressional candidates with poor environmental voting records. Environmental Action, a Washington-based public interest organization, characterized the filthy five as attempting to buy political power with their political contributions. The campaign coordinator of this organization further stated that the filthy five had made a mockery of our nation's pollution control effort by seeking to substitute campaign contributions for clean-up expenditures.

Dow faces an EPA suit, asking for $20 million in fines for chronic air pollution violations at its Midland, Michigan, plant. In 1979 alone, 358 water pollution proceedings were brought against Dow. Since 1973 the company has received 96 OSHA citations. Currently, Dow is being sued by hundreds of victims of its phenoxy herbicide products, such as 2,4,5-T and Agent Orange.

Not too long ago, Dow Chemical Company was a leader among chemical companies in instituting a medical surveillance program to protect workers from potential carcinogenic or mutagenic effects of chemicals. An exemplary program was initiated and refined over a 13-year period to detect chemically induced chromosome abnormalities in workers. By 1974, 43,044 studies were carried out on 1,689 workers involved in the production of various chemicals. During that same period of time 25,000 studies were performed on 1,302 individuals examined prior to their employment. I am pleased to say that Dow established a model program for testing workers for the presence of potential carcinogenic and mutagenic agents. Dow indeed was proud—justifiably—of their accomplishments in this important area of worker protection. In 1977 the scientists at Dow Chemical Company found an increase in cytogenetic abnormalities with exposure to benzene and epichlorohydrin. These studies proved the value of this approach. Shortly after finding that benzene and epichlorohydrin increased chromosomal aberrations in exposed workers, the extensive screening program instituted at the Dow Chemical Company in Texas was phased out. Workers exposed to various agents are no longer being monitored and the effort that was initiated in 1964 with the laudatory goal of identifying carcinogenic and mutagenic agents in the workplace is no longer operational.

I would now like to discuss the area of direct concern to the stockholders of Dow, the agenda items presently under consideration on 2,4,5-T and silvex. I will not cover the extensive literature detailing the hazards of these chemicals. At last year's meeting the hazards of these products both in animal and in limited human studies were described in detail. I would like to report on two very recent studies. The results of these current studies, I am convinced should lead to an overwhelming vote in favor of the proposed resolution. We now have an opportunity for Dow to change its poor image as exemplified in being named to the "filthy five" and seize the initiative in a matter of vital environmental concern.

The first study that I wish to present has to do with an increased risk of cancer and the second study deals with birth abnormalities.

In 1979, Swedish scientists clearly indicated an increase in soft tissue sarcoma among workers exposed to the chlorophenoxy-herbicides and chlorophenols. To estimate whether the excess risk found in the Swedish study can be detected among workers in the United States, Honchar and Halperin reviewed reports of four previous studies including one from Dow to see if an increase in soft-tissue sarcoma was present. The work was published in this year's January issue of Lancet. The workers from the four United States studies were exposed either to 2,4,5-T or to 2,4,5-trichlorophenol. In combining the results of the individual studies, the authors were able to show a marked increase in soft-tissue sarcoma, thus confirming the Swedish work with United States workers.

The number of agents that have been positively identified in epidemiological studies as inducing birth anomalies are indeed few. The drug thalidomide was identified as producing the rare syndrome, phocomelia. With industrial chemicals, seldom have we implicated a specific agent as inducing any of the common birth abnormalities. Given this background of our inability to detect agents that induce common birth defects, one has to be especially concerned when a positive study is reported. On April 17th of this year a report in Science indicated such an effect was found with 2,4,5-T when the rate of birth malformations in the northern region of New Zealand was evaluated. The authors found a statistically significant association between spraying of 2,4,5-T and malformation in the induction of clubfoot. Additionally, a significant increase was found in all birth malformations, especially in heart malformations, hypospadias, and epispadias. When the authors evaluated regions of high, low and intermediate spray density for all birth malformations, the incidence appeared to be positively associated with exposure both across areas and across years.

The evidence from both animal and human data presented at last year's stockholders' meeting was certainly compelling and should have been more than sufficient for a positive response to the resolution. The additional information reported this year should remove any doubt about the potential hazards of these chemicals.

As I understand it, following last year's shareholders' meeting, Dow management responded to the resolution by stating (1) a thousand company studies show 2,4,5-T to be safe, (2) there is not one shred of scientific evidence which questions the safety of 2,4,5-T, (3) all criticisms of herbicides are positioned on nothing more than emotionalism. Given the overwhelming information available and documented at last year's stockholders meeting the response by Dow was indeed ludicrous. Given the information that has surfaced literally within the last few months, for Dow to maintain that position would be inconceivable.

As a scientist who has worked closely in many instances with industry, and specifically with Dow Chemical, I am deeply distressed. Let us start changing this image. With the vast information on the potential and actual harm produced by 2,4,5-T and silvex, it is imperative to support the resolution establishing the review committee to examine and evaluate the existing and potential health consequences of 2,4,5-T, silvex, and their derivatives. How can any logical member of the Dow community not be in favor of this proposed resolution?

References

Environmental Action/News Release April 22, 1981

Honchar, P. A., and W. E. Halperin. Lancet, January 3, 1981, pp. 268-269

Hanify, J. A., et al., Science 212, April 19, 1981, pp. 349-351

Appendix D

NORTHWEST COALITION for ALTERNATIVES to PESTICIDES
BOX 375, EUGENE, OREGON 97440 (503) 344-5044

WARNING!

The following list of pesticides was received from the U. S. Environmental Protection Agency by NCAP in response to a Freedom of Information Act request. Health and safety information used to register these pesticides (as well as others which have not yet been disclosed) was developed by Industrial Bio-Test Laboratories, Inc., the nation's largest independent commercial laboratories. IBT is now under investigation for data falsification. Ninety-nine percent of the cancer studies involved are considered invalid. Other invalid studies were for birth defects, mutations, toxic effects on the nervous system, and tests to determine toxicity of the chemical if swallowed, breathed or absorbed through the skin. EPA has not revealed which studies for what chemicals are invalid. Please copy this warning and pass it on. For further details, contact NCAP.

4-Amino Pyridine – Philips Petroleum
Acarol – Ciba Geigy
Accel – Shell
Adocide–Nalco
Alachlor–Monsanto (Lasso)
Alar–Uniroyal
Alnap–Uniroyal
Altosid-Methoprene–Zoecon
Ametryn–Ciba Geigy
Antor–Hercules
* Asulam–Rhodia
* Atrazine–Shell
Avadex–Monsanto
Avenge–American Cyanamid (difenzo-quat)

Azodrin–Shell
* Bacillus Thuringiensis – Sandoz (Wander)
Barban – Gulf (Carbyne)
Bardike – International Pain Company
Baygon—Chemagro (Proxpur)
Benzadox – Gulf (Topside)
Bicep – Ciba Geigy
Bifenox – Mobil (Modown)
Binapacryl – FMC (Ensodan)
Bis (Tri-M-Butylin) Oxide – M&T
Bladex – Shell (Cyandizine)
Bolero – Chevron (Thiobencarb)
Botran – Upjohn
Brodifacoom-ICI

Bromfenoxim-Ciba Geigy

Bromopropylate—Ciba Geigy

Busan 74—Buckman (TCMTB)

Butam—Gulf

Bux—Chevron

* Cacodylic Acid—Ansul

CA (OCL)$_2$—FMC (Cyanurate)

Captan—American Seed Company/ Chevron

* Carbaryl—Monsanto

Carbofuran—FMC (Furadan)

Cela—Triforine—Chevron

Chipco—Rhodia (lprodione)

Chlorobromuron—Ciba Geigy (Maloren)

Chloropropham—PPG (Furloe)

P-Chlorophenyl-N,N1-dimethyl Allphanate—PPG

Chloropropylate—Ciba Geigy

Chloropyrifos—Velsicol (Dursban)

Chlorothalonil—Diamond Shamrock (Daconil)

Cidial—Montedison (Elsan, Phenthoate)

Ciodrin—Shell (Crotoxyphos)

Cobex—U.S. Borax

Curacon—Ciba Geigy (Profanofas)

Cycle—Ciba Geigy (Procyanazine)

Cycocel—American Cyanamid

Cyprazine—Gulf (Outfox)

Cypromid—Gulf

*2,4,—D—Hercules/Dow (NACA)

Daconil—Diamond Shamrock

Dacthal—Diamond Shamrock (DCPA)

Dasanit—Chemagro (Fensolfothion)

D.C. 5700—Dow Corning

DDVP—Zoecon

Delachlor

Delnav—Hercules (Dioxathion)

Desmedipham—Nor Am

*Dicamba—Velsicol

Dichlobenil—Casoran

Diethanol Amine 2,4-Di-Chlorophenoxy Acetate—Chevron

Dietbanol Amine 2-(2-Methyl-4-Chloro-phenoxy) Propionate—Chevron

Difolatin—Chevron (Captafol)

*Dinoseb—Dow

Diquat—Chevron

Di-Syston—Chemagro (Disulfoton)

DMDMH—55—Glyco (Dantoin)

Drepaman—Montedison

DSMA—Diamond Shamrock

Dual—Ciba Geigy (Metolachlor)

Dynap—Uniroyal

Embark—3M (Mefluidide)

Endothall—3M/Pennwalt

EPN—E.I. Dupont/Velsicol/Nissan

Ethalfuralin

Ethiolate—Gulf

Fenitrothion—Sumitomo (Sumithion)

Fenvalerate—Shell

Flo Mo—Sellers

Fluoridam—3M

Fluorodifen—Ciba Geigy (Preforan)

Folpet—Chevron/Stauffer (Phaltan)

Formetanate

Formetanate Hydrochloride—Nor Am (Carzol)

Gossyplure—Conrel (Nomate)

Glutaraldehyde—3M (Arbrook)

* Glyphosate—Monsanto (Roundup)

Glyphosine—Monsanto (Polaris)

Harvade—Uniroyal

Hexakis

Hinosan—Chemagro (Edifenphos)

Iodin—PPG/AJAY

Iodrin—Shell

Iprodione—Rhodia (Chipco)

Irgasan—Ciba Geigy (Trichlosan)

* Krenite—E.I. Dupont

Lanstan—FMC

Machete—Monsanto (Butachlor)

Maleic Hydrazide—Uniroyal

Malonoben—Gulf

*MCPA—Diamond Shamrock

MCPP—Rhodia

Merphos—Mobil

Mesurol—Chemagro (Methiocarb)

Metabromuron—Ciba Geigy

Metasistox—Chemagro

Methazole—Velsicol

Methomyl—E.l. Dupont (Lannate)/Shell (Nudrin)

Meteran—FMC (Polyram)

Methoprene—Zoecon

Metolachlor—Ciba Geigy (Dual)

Metribuzin—Chemagro (Sencor)

Mocap—Mobil (Ethoprop)

Monitor—Chevron (Methamidiphos)

Monochlorotoluene

*MSMA—Diamond Shamrock

Nalco 247—Nalco

Nalco 341—Nalco

Nalco 2210—Nalco

Nalco 7325—Nalco

Nalco 7340—Nalco

Nalco 7623—Nalco

Naleo 7644—Nalco

Nalco 8844—Nalco

Nalco 4WC 317—Nalco

Naled—Chevron (Dibrom)

NaO CI—Jones Chemical/ P&G

Nemacur—Chemagro (Fenamiphos)

Nemagon—Shell (DBCP)

Nemaphene—Shell

Nicotine/Nicotine Sulfate—Black Leaf

Norea—Hercules (Herban)

ODAA—Arbrook (Glutaraldehyde)

Omadine—Olin

Omitt (Comite)—Uniroyal

* Orthene—Chevron

Oxadiazon—Rhodia

Oxycarboxin—Uniroyal

Paraquat—Chevron

Pencap E—Pennwalt

Pencap M—Pennwalt

Perfluidone—3M

Permethryn—FMC

Phenmedipham—Nor Am (Bentanal)

Phorate—American Cyanamid

Phosalone—Rhodia

Phosphamidon—Chevron (Dimecron)

* Picloram—Dow (Tordon)

Pik—Off—Ciba Geigy

Piperonyl Butoxide—MGK

Plictram—Shell

Polyram—FMC (Metiram)

Potassium Hexafluorarsenate—Pennwalt

Potassium and Sodium Azide—PPG

PPG 124—PPG

PPG 135—PPG

PPG 148—PPG

PPG 171—PPG

PPG I72—PPG

Profluralin—Ciba Geigy

Promecarb

Prometon—Ciba Geigy (Promitol)

Propanil

Propham—PPG

Prowl—American Cyanamid

Pydrin—Shell

Pyrethrin—MGK

R-69—S.C. Johnson

Rabon—Shell (Gardona)

Ramrod—Monsanto

Randox—Monsanto (Allidichlor)

Rydex—U. S. Borax

Santophen—Monsanto

Sectrol—3M

Sencor—Chemagro (Metribuzin)

* Silvex—Rhodia

* Simazine—Ciba Geigy

Sodium Azide—PPG

Sodium Bromide and Tri-chloro-s-Tri-azine-Trione—Jones

Sodium Chlorate—Pennwalt

Sumitol—Ciba Geigy

Sumethryn

Supracide—Ciba Geigy (Methidathion)

System E—3M

System M—3M

5-2,3,3-Trichloroallyl diisopropyl Thioca-neanate

Terbufos—American Cyanamid

Terbutbylazine—Ciba Geigy

Terbutryn—Ciba Geigy

Terrazole—Olin Corporation

Thidiazuron—Nor Am

Thiodan—FMC (Endosulfan)

Thiofanoz—Diamond Shamrock

* Thiram

TL 1076—MGK

Topsin M—Pennwalt (Thiophinate)

Torak—Hercules (Dialifor)

Toxaphene—Hercules

Triadine—Olin

Triforine—Chevron

Triphenyltin Hydroxide—Thompson-Hay-ward

Trivax—Brandington

Vapona—Shell (DDVP)

Vegadex—Monsanto (Sulfallate)

Vel—Sandoz (Propetamphos)

Vendex—Shell

Valpar—Dupont

Vinyzene—Ventron

Visco Adomall—Nalco

Visco P—25—F—Nalco

Visco 1151—Nalco

Visco 1152—Nalco

Visco 1153—Nalco

Visco 3990—Nalco

Vitavax—Uniroyal

Vydate—E.I. Dupont (Oxamyl)

Zinophos—Nor Am

*Chemicals commonly used in forest management in the Northwest.

Appendix E

Resolution on Agent Orange, National Council of the Churches of Christ in the U.S.A. Adopted by the Governing Board November 9, 1979

Whereas there is substantial indication that certain phenoxy herbicides (2,4-D and 2,4,5-T) and a contaminant, dioxin, widely used in the Vietnam conflict under the name "Agent Orange,'" may be the cause of a wide range of serious physical and psychological ailments among human beings subjected to them; and

Whereas many U.S. veterans of the Vietnam conflict (not to mention Vietnamese people and Australian, Korean and other allies of the U.S.) are suffering from serious chronic symptoms attributed to contact with herbicides in Vietnam; and

Whereas many other people in the U.S. (and elsewhere) have suffered similar disabilities from contact with civilian use of such herbicides in industry and agriculture, so that the Environmental Protection Agency has imposed a (partial and temporary) ban upon their use; and

Whereas the Veterans Administration has resisted recognizing the disabilities attributed to contact with herbicides as "service-connected" and thus provides no medical or other help to veterans suffering such disabilities, yet has found no other explanation for them; and

Whereas the manufacturers of these herbicides are pressing for the lifting of the EPA ban so that they can continue their manufacture, sale and use; and

Whereas the burden of proving the harmful effects of herbicide exposure presently rests almost exclusively on those who are or could be victims of its use, which burden of proving safety more properly rests on the users and producers of such substances; and

Whereas presently the entire burden of proof of injury from the hazardous chemical "Agent Orange" rests on the claimant;

Therefore, be it resolved: that the Governing Board of the National Council of the Churches of Christ in the U.S.A.:

1. Urges its member churches to study the current crisis related to "Agent

Orange" and other toxic herbicides used in the Vietnam War and also in ongoing domestic defoliation;

2. Calls on its member churches to examine their corporate investments in companies that produce products containing dioxin, and to take appropriate stockholder actions;

3. Calls to the attention of its member churches the formation of the new National Veterans' Task Force on "Agent Orange,"' including the National Conference on "Agent Orange" scheduled for November 11-13th in Washington D.C.;

4. Calls on the religious community as a whole to join with U.S. veterans and others in the U.S. and the international community in raising the moral and ethical issues related to all those affected by the production and use of toxic herbicides, including the lives and land of the Vietnamese, and the health of many of those who participated in the U.S. military efforts, including U.S. civilians and the military forces of Australians, Koreans and other U.S. allies in Vietnam;

5. Calls for a complete and comprehensive moratorium on the use or export of these toxic herbicides and related chemicals until it can be established without reasonable doubt that negative effects do not outweigh benefits;

6. Calls for both government and independent studies to determine the symptoms of, and possible treatments for, the effects these chemicals have already had on the persons and environments exposed to them;

7. Calls on the scientific community, along with the government agencies and corporations involved, to exercise their moral responsibility as well as their scientific acumen in conducting these studies in a factual, balanced and objective manner;

8. Calls for full disclosure by those who conduct studies or possess other relevant knowledge, in order to facilitate corrective action including but not limited to, medical treatment of persons affected and to assist and ensure full public debate on this subject; and

9. Calls for the government to require itself or the responsible corporation to carry the burden of negating any claim that illness or injury resulted from exposure to "Agent Orange".

Based on Policy Statement: "Consumer Rights and Corporate Responsibility" February 14, 1972.

"Christian Principles and Assumptions for Economic Life"—September 15, 1954. "The Ethical Implications of Energy Production and Use"—May 11, 1979. "Christian Concern and Responsibility for Economic Life in a Rapidly Changing Technological Society"—February 24, 1966.

Appendix F

Resource Organizations	
Northwest Coalition for Alternatives to Pesticides P.O. Box 375 Eugene, Oregon 97440 (503) 344—5044	National Council of Churches 475 Riverside Drive New York, N.Y. 10027 (212) 870—3600
Friends of the Earth 530 7th Street S.E. Washington, D.C. 20003 (202) 543—4313	Interfaith Center on Corporate Responsibility 475 Riverside Drive, Room 566 New York, N.Y. 10027 (212) 870—2293
Southern Coalition for the Environment P.O. Box 3021 Hammond, Louisiana 70404 (504) 345—4339	Citizen Soldier 175 Fifth Avenue New York, N.Y. 10010 (212) 777—3470
National Veterans Task Force on Agent Orange National Veterans Law Center 4900 Massachusetts Avenue N.W Washington, D.C. 20016 (202) 686—2741	National Association of Farmworker Organizations 1329 E Street N.W, Suite 1145 Washington, D.C. 20004 1—800—424—5100 (24—hour emergency hotline for farm—worker poisoning cases)

Appendix G

In the spring of 1980, while many of the events described in *A Bitter Fog* were unfolding in Oregon, more than 1300 miles away in Canada an investigative reporter was painstakingly unearthing a story that would directly impact those events. Peter von Stackelberg, a reporter with *The Leader-Post* in Regina, Saskatchewan got a lead that would over the next year result in more than 50 stories about problems with the safety of hundreds of drugs, pesticides, and industrial chemicals.

The importance of Peter's lengthy series of investigative reports into errors, negligence, and out-and-out fraud in testing the safety of chemicals and drugs was recognized with several awards.

In 1981, the Governor General of Canada presented Peter with the Citation of Merit, Roland Michener Award for Meritorious Public Service in Canadian Journalism for investigative reporting on science, technology, and environmental issues done while working with The Regina Leader-Post. The Michener Award is one of the highest distinctions in Canadian journalism and is presented to a Canadian news organization "whose entry is judged to have made a significant impact on public policy or on the lives of Canadians".

That same year Canada's Centre for Investigative Journalism (CIJ) published one of the stories on problems in the testing of chemicals and drugs in its 1981 Annual Review, which featured some of Canada's best investigative reporting.

In 1988, Sonoma State University's "Project Censored" cited his work on environmental issues as number two of its 25 most important "censored" stories of the year. Each year the Project Censored researches, vets, and compiles the Top Twenty Five most censored and under-reported news stories in the US, and offers scholarly analysis and critiques, and publishes them.

A selection of those stories are provided in this appendix.

* * *

Original Foreword to The Toxic Mist

The use and safety of chemical pesticides promises to be one of the major environmental issues of the 1980s.

Increasing food production to feed more people on this planet will increase the pressure on farmers to use more and more chemicals. At the same time, pressure from scientists, environmentalists, and consumers could mean government action to restrict the use of many chemicals that are now on the market.

This series of stories started in June, 1980 when rumors that the Saskatchewan government was planning to ban a number of chemicals reached The Leader-Post. While the rumors were wrong, they prompted The Leader-Post to assign reporter Peter von Stackelberg to do some research into the testing of pesticides and the problems government was facing because of the Industrial Biotest Laboratories (IBT) affair.

Since June The Leader-Post has published more than 50 stories about the use and safety of chemical pesticides. Many of the major stories from among those 50 are reprinted in this booklet.

Peter von Stackelberg joined The Leader-Post as a reporter in October, 1979. Prior to working on the agriculture beat and chemical pesticide stories he covered the Saskatchewan Legislature.

Before joining The Leader-Post, he worked as a reporter with The Daily Herald Tribune in Grande Prairie, Alta. During the four years he was there, von Stackelberg covered agriculture, environmental issues in forestry, the city hall beat and a number of other areas.

He is a graduate of the journalism program at Ryerson Polytechnical Institute in Toronto,

He was born on Oct. 28, 1953 in Elk Point, Alta. (because it had the closest hospital to the farm near Maidstone, Sask. where his parents lived). His family settled in Peace River, Alta. in the late 1950s.

He grew up in Peace River and in the late 1970s helped clear and work homestead land just north of Peace River.

* * *

Introduction to The Toxic Mist

A lot has been said in recent years about the limitations on this country's economic and political independence.

But what price are we prepared to pay for our chemical sovereignty?

To put it bluntly, Canada has been freeloading when it comes to the data government uses to decide the acceptability and safety of chemicals and drugs used in industry, the home and on the farm.

We have also relied heavily on foreign countries, mainly the U.S., for disposal of toxic materials, notably polychlorinated biphenyls (PCBs).

The time may have come to stand back and examine the role chemicals are playing in our lives – and possibly our deaths. What are the costs of continued or expanded use of chemicals? What are the costs of curbing their applications?

The appropriate vehicle for examining such questions might be a federal royal commission.

The dangers of relying on foreign findings were underlined in the case of Industrial Biotest Laboratories, the private Illinois testing facility which faked results of close to 100 pesticides.

There is the case of TOK, the herbicide which was recently banned in Saskatchewan. TOK testing, which later proved inadequate, originally took place in the U.S. and Europe. It was this information, supplied by the product's producer, which Health and Welfare, and Agriculture Canada drew on in earlier approving TOR under the Pest Control Products Act. (It was also new foreign evidence which led to TOK's withdrawal)

TOR also illustrates the divided responsibility not only between levels of government but also within the same jurisdiction on testing, licensing, regulating, labelling, marketing, transporting, using, and disposing of chemicals. This is true not only of pesticides but also of drugs, food additives and so-called industrial chemicals.

Health and Welfare Canada recommended last October that TOK, understood to cause birth defects, be withdrawn. However, Ottawa agreed to allow the manufacturer to keep marketing existing stocks.

Defining the lines of jurisdiction within and among governments is at times as difficult as tracing what happens to the chemicals themselves.

For instance, while it is Agriculture Canada which licenses pesticides, it does so after Health and Welfare has reviewed the manufacturer-supplied data on harmful effects of chemical residues in crops, as well as the health impact on those in contact with the pesticide.

Health and Welfare, which also has to approve new types and uses of drugs and food additives, admits practically all the testing on drugs it is asked to approve is done by foreign laboratories. Canada has limited facilities and the costs of in-country testing would be prohibitive in a "small" country like Canada, according to Health and Welfare Minister Monique Begin.

Agriculture Canada does test out pesticides under Canadian conditions but also has relied mainly on data generated elsewhere to decide the safety of approximately 430 pesticides on its approved list.

While there at least is a formal licensing process for farm-related chemicals and pharmaceuticals, there isn't for so-called industrial chemicals. The U.S. has established a list of about 30,000 such substances. That list is expanding by about 800 new products a year.

Environment Canada can act under the Environmental Contaminants Act to place materials on its prescribed list, but generally only does so after a product is introduced on the market.

Other federal departments come into the chemical equation on the transportation and labelling of chemicals and food additives.

Provincially, while the department of the environment has an over-all protective role, it was the Saskatchewan Agriculture-administered, provincial Pest Control Products Act which was applied in the TOK case. The federal and provincial departments of labor, through their occupational health and safety branches, are to ensure chemicals are properly used and stored at the work place.

We are not for a minute suggesting a royal commission is going to untangle all these crossed and overlapping lines of authority. However, royal commissions have been favored means to review issues of vital and lasting national concern. The roles man-made chemicals play in our lives surely fit into this category. Such an inquiry need not be a witch-hunt to tick off thousands of chemicals as culprits.

* * *

Not Proven Safe

Fraudulent or inaccurate test results for more than 100 pesticides, many used in Saskatchewan, have raised concerns about their long-term hazards.

A letter obtained by *The Leader-Post* says the results of long-term studies of the chemicals "are considered invalid" by the federal health and welfare department.

"This means that a number of chemicals in use in Saskatchewan have not been proven safe.

The letter, written Jan. 25,1980, by R.O. Read, chairman of a federal committee on pesticides, said the validity of all studies carried out by Industrial Biotest Laboratories (IBT) of Northbrook, Ill. "remains in doubt."

"All long-term rodent studies and multigenerational reproductive studies performed by IBT are considered invalid," Read said.

Other studies would have to be used to show the long-term safety of the chemicals.

In his letter, Read noted that government authorities were forced to send letters to a number of chemical companies, pointing out that many had "failed to submit the information requested by the Environmental Protection Agency and the Canadian health protection branch."

He also said spot checks by government officials had "uncovered serious problems which were not identified or reported" by chemical companies reviewing IBT test data.

"Most of the critical studies that have been audited are not satisfactory."

Read said the government's policy on the matter is that "the validity of all IBT studies remains in doubt until successfully demonstrated by the sponsoring company to be otherwise."

Questions about a wide variety of chemicals were raised about three years ago when an investigation in the United States revealed Industrial Biotest Laboratories had faked the results of some tests.

Sloppy or inadequate records invalidated the results of other tests, according to the investigation by U. S. federal agencies including the justice department and the Environmental Protection Agency.

In some cases, sick test animals were apparently replaced with healthy ones.

The results of the IBT tests were used by regulatory agencies in several countries to establish safe uses for certain pesticides, food additives, packaging materials and drugs.

The Canadian government formed an interdepartmental committee – consisting of representatives from the environment, health and agriculture departments – to examine the matter more than two years ago.

But no list of the chemicals in question has ever been made public by the government.

Sources in the federal government have indicated 96 chemicals are under scrutiny, while a spokesman for the American Environmental Protection Agency said it has 123 chemicals on its list.

The Saskatchewan government has been attempting – unsuccessfully – for the past several months to find out from federal authorities exactly which chemicals are on the list.

Environment Minister Ted Bowerman said he has attempted to get information about the chemicals from the federal government, but has not received any replies.

"There is a major concern. We have a concern over what appears to be a weakness in the testing related to the IBT affair," Bowerman said.

Dr. David Penman, health consultant with the provincial environment department, said not knowing what chemicals might present long-term health hazards leaves the province in a difficult position.

"There are concerns about a number of these chemicals. There is no doubt about that," Penman said.

"If Chemical X is on the list, and is for some reason suspect, it would be a real concern to governments in Canada to know the reason for the lack of validity of the informational base.

"If there are substantial doubts about the reasons why this Chemical X is used, then why is the licensing authority (the federal government) permitting its use!"

Penman said the identification of the chemicals is important.

"We want to be assured that if these agents are used in Saskatchewan, that they are safe for the workers who pack them, safe for the farmers who use them, that there is no hazard in their general use and storage, and that there is no hazard to the public when the chemicals are used and dispersed in the environment."

Bowerman said his department is dependent upon the federal government's findings.

"While at the present time there appears to be a question of whether there has been adequate and sufficient testing done on a number of chemicals, I'm not sure that we are in a position to be able to clearly state that the evidence is such that it would condemn them," Bowerman said.

"So one would have to wait on what the federal government says, and if it's found that (the chemicals) have not been adequately tested, then maybe the thing to do is go back and test them."

Bowerman said it is still too early to speculate about a ban on the use of the chemicals in question.

But he suggested that withdrawal of federal approval until testing is redone might be appropriate.

"If we were to get the response (from the federal government), the opinion definitely expressed that there was inadequate laboratory testing, inadequate results or that the results were somehow jiggered or misrepresented, then one would have to seriously consider a temporary ban."

Sources within the federal government said information about the chemicals was to have been forwarded to provincial authorities earlier this year, but senior government officials stepped in and prevented that from happening. According to those sources, a letter outlining the situation was prepared and forwarded to federal Health and Welfare Minister Monique Begin for her signature some time ago.

But the letter has not yet been received by officials of Saskatchewan's environment department.

The Leader-Post has obtained a partial list of the chemicals in question. Among them are:

- Captan – a fungicide used in seed treatment and in garden and greenhouse crops

- Carbonfuran – used as an insecticide, particularly on flea beetles and grasshoppers, being used in Saskatchewan this year to combat outbreaks of those insects
- Diazanon – an insecticide that is widely used on gardens and can be obtained at most garden supply stores

American studies have created concern about captan's role in causing cancer. In a 1974 study Dr. Marvin Anderson of Henry Ford Hospital in Detroit and Dr. Herbert Rosenkranz of Columbia University said captan "is highly suspect to be a carcinogen."

The doubts raised about the validity of long-term studies of the chemicals tested by IBT do not mean there is necessarily a danger.

But it does mean that chemicals that were thought to be safe might not necessarily be so.

* * *

Pesticide Controls – Are They Failing?

For years we have assumed the pesticides we use in our homes, gardens and on farms are safe.

But are they?

Fraud and incompetence at laboratories testing multitudes of chemicals have raised major concerns about their safety.

The regulatory system for pesticides appears to be failing.

It has been condemned as having "major deficiencies" by the Saskatchewan environmental advisory council.

Farmers, grain elevator operators, and others suffer from health problems related to chemical exposure.

Yet chemicals whose safety is in doubt continue to be widely used.

Residues of these chemicals contaminate food and pollute lakes, rivers, wells and other sources of water.

A stinging condemnation of the regulatory system for pesticides is contained in the 1977-78 and 1978-79 annual reports of the Saskatchewan government's environmental advisory council.

"There are major deficiencies in the present research and regulatory process," the council said in 1978.

It said the federal government does not do enough independent research on the chemicals it approves.

"Both (the agriculture and health and welfare) departments are forced to rely in part on laboratory tests by the chemical manufacturers."

The advisory council said the lack of objectivity and credibility are sometimes missing when tests are sponsored by manufacturers.

The scandal involving Industrial Biotest Laboratories of Northbrook, Illinois is just one example of this.

Questions about the safety of about 100 pesticides was raised in 1977 when an investigation in the United States revealed IBT had faked the results of some tests.

In some cases sick test animals were apparently replaced with healthy ones Sloppy or inadequate records invalidated the results of other tests, according to U.S. federal agencies investigating the fraud. According to critics, this fraud, sloppiness or incompetence can be the result of pressure – subconscious or deliberate – on scientific staff with the company by research and development or marketing departments in a hurry to get the chemical onto the market and start making money.

Critics suggest independent laboratories, eager for lucrative contracts in the future, might also be influenced into producing information or interpretations of data that are thought to be in the interests of the company they are testing products for.

"A proliferation of many commonly used chemicals has occurred without sufficient prior examination of the potential consequences. These have instead, become known in a costly and, in some cases, tragic manner," the advisory council said.

The council said comprehensive study of the thousands of commonly used chemicals for harmful effects has not kept pace with the rate at which they are introduced.

"The fact remains that most of the chemicals in common use have yet to be tested at all."

The advisory council was also critical of the lack of information about how many toxic chemicals are in use in Saskatchewan, the quantities of chemicals used, and where they are sprayed.

"On this basis alone existing provincial control mechanisms for the regulation of toxic chemicals must be deemed inadequate."

While the IBT affair is the most glaring example of fraud and incompetence, there are others.

Samuel Epstein, professor of occupational safety and environmental medicine at the University of Illinois, said in the December, 1979 issue of The Ecologist there are a number of other cases of deliberate manipulation of tests or suppression of important information.

In an article entitled Polluted Data Epstein said fraudulent manipulation of data has been found for:

- The drug MER/29, resulting in the criminal conviction of several officials of Richardson-Merrill Company.

- The chemical Dornwall produced by the Wallace and Tiernan Company.
- McNiel Laboratories pleaded no contest on charges of concealing information about the chemical Flexin.
- Hazelton Laboratories was charged with falsifying data on the artificial sweetener aspartame and the drug Aldactone.
- The drug Panalba was withdrawn from the market in 1968 after information in the manufacturer's files was accidentally discovered by an American Food and Drug Administration inspector.
- The Velsicol Chemical Company was indicted by a U.S. grand jury in 1977 for suppressing information that Chlordane/Heptachlor was carcinogenic.

But the IBT affair has had the most serious effect.

The Canadian government has determined more than half of the 405 1BT studies examined up to March 28, 1980 were of no use in determining the safety of a variety of chemicals approved for use in Canada.

A further 410 IBT tests have yet to be examined to find out if they are reliable. Although the safety of 97 chemicals on the Canadian list could not be guaranteed, the federal government permitted their use during the past three years.

It will be at least another year before all the tests done by IBT will have been examined to find out if they were tampered with. The federal government is now facing demands from the Saskatchewan environment department and from members of Parliament that it release information about why 233 tests done on a number of chemicals on the list of 97 were invalid.

Dr. David Penman, health consultant with the provincial environment department, said it makes a big difference if the tests were invalid because relatively minor data about the melting point of a chemical was improperly tested or if the very important tests establishing the cancer-causing properties of the chemical were wrong.

The provincial government will continue to push for release of all information regarding the IBT tests, environment minister Ted Bowerman said.

"If I do not receive more information promptly of if I am not satisfied that federal officials are taking appropriate measures to re-establish the status of the chemicals in question, I will be consulting with my colleagues in provincial health and agriculture to consider what measures Saskatchewan should take on its own," Bowerman said earlier this week.

* * *

U.S. May Upset Registrations Here

The Canadian system for registration of chemicals could be thrown into turmoil by events in the United States.

Dr. Alex Morrison, federal deputy health minister, said the chemical industry might find significant changes forced on it if American investigations produce evidence of widespread fraud or incompetence in the testing of pesticides, drugs, food additives, and industrial chemicals.

Fraudulent and inadequate testing by Industrial Biotest Laboratories (IBT) of Northbrook, Ill. cast doubts upon the safety of the more than 125 compounds, including pesticides, food additives and drugs in commercial, agricultural and domestic use in Canada. Major problems with laboratory work carried out by IBT were discovered by American authorities in 1977.

"If IBT is an aberration we can deal with it," Morrison said.

"But a widespread inability to trust them (laboratories and chemical companies) could mean the government stepping in."

Morrison said his department officials have "heard rumors but have not yet been able to verify" reports of the U.S. Environmental Protection Agency investigating a number of firms because of inadequate or fraudulent testing of chemicals.

Earlier this week the Leader-Post revealed the EPA was investigating, in the words of an EPA official in Washington, "nine or 10 laboratories because of the possibility of careless or deliberately careless work."

Morrison said he would prefer to see the primary responsibility for testing of chemical products remain in the hands of the manufacturers. The government's role would be mainly that of auditing the data presented by the manufacturers.

He said the laboratory testing work carried out by government would-be of a limited nature, enough to give federal officials the competence necessary to thoroughly check industry data.

Morrison said it would be beyond the capability of the Canadian government to get involved in the actual testing of chemicals.

The financial burden – about $1 million per chemical – would be excessive if the federal government had to test each of' the hundreds chemicals put on the market every year. he said.

The possibility of government, which regulates the use of chemicals, getting into testing could also lead to conflicts of interest, Morrison said.

He said government scientists would also be subject to the same temptations that faced researchers with IBT.

"You don't want to trust government to do things totally, any more than you would trust industry," Morrison said.

He said the best testing and regulatory system would involve a mix of government, private industry and international agencies such as the Food and Agricultural Organization (FAO) and World Health Organization (WHO) of the United Nations.

That is essentially the system now in place, critics say. Simon de Jong, MP for Regina East, said it is obvious that Canada needs its own testing facilities.

He said a number of private laboratories, like IBT, have shown they are simply not morally capable of performing the tests needed to ensure the long-term safety of chemicals.

De Jong was critical of statements made last month by federal Health Minister Monique Begin that Canada is a small country that would have trouble testing all the pesticides in use here and therefore should take advantage of information coming from other countries, particularly the U.S.

"The information we have been obtaining from other countries has all proven to be false.

"The argument for our own independent, trustworthy labs is so common sense that their rejecting it makes me suspicious."

Dr. David Penman, senior health consultant with the provincial environment department, said the IBT affair shows there is a major defect in the Canadian regulatory system.

He said it is illogical that Canada, as a major industrialized nation and one of the largest producers of food, has no independent ability to carry out testing of pesticides used on food crops. "When people say we can't have this testing capacity I challenge that."

He said Canada already has the ability to carry out major studies of prescription drugs.

"Many of the major chemicals used in agriculture in Canada are implicated In this IBT affair.

"It is important that we have our own national capacity, under the direction of Canadians and serving the needs of Canadian agriculture."

The Saskatchewan government continues to push for more information on how the tests for the chemicals under suspicion were flawed.

Penman says it is essential for the province to know what data was faked and improperly gathered so those exposed to the suspect chemicals can take effective measures to minimize the health hazards.

Morrison told The Leader-Post Friday that the IBT tests in question are mainly those into the long-term health hazards of the chemicals.

While the provincial government continues to make a fuss about getting more information from the federal government, officials in Ontario seem to be less concerned.

Dr. George Cooper, chairman of the Ontario government's pesticide advisory committee, downplayed the effect of the IBT scandal. "We have known all about this for the last three years," he said.

But the committee has not made any recommendations to the Ontario government about it, Cooper said.

"We don't really feel it is our job to advise the minister (of environment for Ontario) as to what to do about the IBT affair. We do recommend on many things, but not on this matter. We feel it is a federal matter, rather than a provincial one."

While provincial governments may have different views on their responsibilities, the costs of the IBT affair will touch everyone.

Those costs are colossal.

The Canadian government has already spent more than $600,000, simply reviewing the information published by IBT, Morrison said.

But that pales beside the costs the chemical companies wlll have to pay for retesting the chemicals involved.

About $100 million will have to go into this work, Morrison said. And the costs could be much greater.

Suddenly banning the use of the chemicals in question would create massive problems in agricultural production and the food processing industry.

If the scandal of fraudulent and improper testing spreads even further, and involves numerous other firms, the economic effects could be staggering.

* * *

Hazardous Farm Chemical Still Available Across the Country

A dangerous agricultural herbicide is being sold almost a year after federal health officials recommended banning it because there was no efficient way of stopping its sale sooner.

Wayne Ormrod, associate director of the pesticides section of the federal agriculture department, said Friday the voluntary withdrawal of the herbicide TOK by its manufacturer was the most effective way of dealing with the matter,

"If more formal regulatory action was taken against the registration, and there was an appeal, it could have taken much longer to remove it from the market. In many cases like this a voluntary withdrawal by the company is the most efficient," Ormrod said,

The chemical TOK – sold under the trade names TOK G-25 and TOKIRM Selective Herbicide – is known to cause birth defects. The herbicide is used on rapeseed crops on the Prairies and on some commercial vegetable crops, primarily onions, in other parts of the country.

U.S. plans ban

The American Environmental Protection Agency (EPA) is planning to impose a ban on the manufacture and sale of the chemical in the United States within the next two weeks, an official in Washington told The Leader-Post Friday. He asked that his name not be used.

Evidence of the potency of TOK in causing birth defects led California to put severe restrictions on its use, the official said.

"One dose of it can cause birth defects," he said,

"Pregnant women are forbidden to work in fields (in California) where TOK has been sprayed. As well, there is a 14-day waiting period before any workers are allowed in treated fields."

Even then the workers are required to wear rubber gloves, boots and clothing and respirators, he said.

Carol Peacock, an information officer with the health department in Ottawa, said the manufacturer had voluntarily pulled the chemical from the market almost a year ago, after the department recommended it be withdrawn.

But TOK is still available to farmers across the country.

Jack Elliott, marketing manager for Rohm and Haas, said in an interview from Toronto the federal government has allowed the company to sell off its remaining stocks of the chemical.

Eliott and government officials did not have any information about how much of the chemical had been sold this year or how much remained to be sold.

The herbicide was first registered for use on vegetable crops in 1967 and on rapeseed crops in 1970.

At the time scientific information collected by the company indicated the chemical was acceptable for use, Elliott said.

"But more recent studies indicated it was absorbed more rapidly through the skin than originally thought. That reduced the safety factor of the chemical."

Ms. Peacock said the health department's recommendation that TOK be banned was based on information Rohm and Haas submitted last fall.

"The new information indicated it caused birth defects. The second study (which uncovered the birth defects) was done by a different method than the initial one," she said.

Causes cancer

She said studies done by the American National Cancer Institute also indicated the chemical caused cancer. But the federal health department did not go into studying in depth the question of cancer because there was sufficient evidence of birth defects to call for removing the chemical from use.

While the health department can call for the withdrawal of hazardous chemicals from the market, only the federal agriculture department has the authority to pull it.

Peacock said her department went as far as it could and the decision to allow Rohm and Haas to sell off its remaining stocks was made by the agriculture department.

* * *

TOK to be Banned

A ban on the sale and use of the herbicide TOK was announced Friday by the Saskatchewan government.

"TOK is a serious health hazard and continued use of the herbicide cannot be recommended under any circumstances," Environment Minister Ted Bowerman said.

"The Saskatchewan government warns the public, and particularly farmers, that the herbicide is a serious danger. We are moving to place a prohibition on its use and sale," Bowerman said.

The Leader-Post revealed last week that scientific data presented to the federal government last October showed the chemical to be highly dangerous. Extremely low doses given to mice in laboratory tests caused deformation of the diaphragm muscle – essential to breathing – to such an extent many of the animals died shortly after birth from suffocation. Another problem discovered in test animals was that glands similar to the human tear glands failed to develop.

The chemical is absorbed through the skin as well as entering the body through the digestive and respiratory systems.

Bowerman said Saskatchewan residents who have unopened cans of the chemical – sold under the trade names TOK E-25 and TOK RM Selective Herbicide – should return them to the dealer.

"Dealers are in turn advised to return the chemical to the manufacturer," Bowerman said.

The herbicide should not be dumped on land or in water, he warned.

"Partially used cans should be securely sealed and enclosed in double plastic bagging. Local agricultural representatives will be able to advise farmers about disposal of the partially used cans of TOK," Bowerman said.

* * *

Widely Used Insecticide Under Attack by American EPA

An insecticide used widely in seed treatments and household products available in Saskatchewan is under attack by the American Environmental Protection Agency (EPA).

The chemical Lindane is also used in a more than 80 different products. Among those registered in the federal government's 1979 listing of pest control products are pet flea and tick collars and sprays, livestock insecticide treatments, and louse and flea preparations for human use.

Use of the insecticide will probably be severely restricted in the United States within the next year, Jim Sibbison, an EPA spokesman in Washington, said.

An EPA document, dated June of this year, obtained by The Leader-Post recommends banning Lindane for almost all uses, including seed treatments, household uses, and for treatment of logs and lumber. The few uses that should be continued would be under severe restrictions. Only licenced applicators would be allowed to work with the chemical, and then only when completely dressed in protective clothing, neoprene aprons, boots, and elbow-length gloves. Special warnings recommending pregnant women avoid exposure to the chemical were also suggested by the EP A.

Lindane is known to cause cancer and other tumors, birth defects, and reproductive disorders in test animals. There have been numerous reports of the chemical causing anemia and several scientific reports have noted Lindane might cause leukemia. The EPA also expressed concern about the toxic effects of Lindane on young children.

"A number of cases of Lindane poisoning in children heightens the agency's concern about the sensitivity of children to the effects of Lindane."

The Saskatchewan environment department just recently started its own examination of the chemical.

The department is attempting to get access to the federal agriculture department documents containing scientific information on the health and biological effects of the chemical, Dr. David Penman, senior health consultant for the environment department, said.

The chemical has major significance for agriculture in Saskatchewan. Lindane is in all the dual-purpose seed treatments listed in the provincial agriculture department's most recent pamphlet on available seed treatment products.

But there are a number of problems with the chemical.

The Saskatchewan government is currently trying to figure out how to get rid of about 500,000 pounds of rapeseed treated with Vitavax – a Uniroyal product that contains Lindane. The germination of the rapeseed was substantially reduced by the seed treatment and seedlings that did emerge were weakened and unable to survive adverse weather conditions.

The chemical was also involved in the death of a Manitoba farmer earlier this year. The death was because of kidney failure caused by exposure to Vitavax, the medical examiner involved in the case said.

Most Saskatchewan residents, whether they use Lindane or not, are exposed to it. The 1978-79 annual report of the Saskatchewan Environmental Advisory Council said Lindane was widespread in rivers, lakes and other surface water in the province. The highest concentrations of Lindane in surface water were found in the Qu'Appelle River district, the advisory council said.

It said most agricultural chemicals tend to accumulate' in sediment on the bottom of lakes and streams.

"Bottom sediments therefore serve as a sink for these toxic substances and. through the process of biological (accumulation), pose not just an immediate but also a long-term potential threat to the environment the council said.

Vitavax is among more than 100 chemicals involved in fraudulent and inadequate safety tests carried out by Industrial Biotest Laboratories (IBT) in the United States.

The chemical remains on the market in spite of assurances by federal Health Minister Monique Begin that doubts about safety would prompt quick government action.

"My officials have been instructed to examine laboratory data with particular reference to detecting any previously reported adverse effects," Begin said July 31 of this year.

"Should any such effects be noted, suggesting a potential health hazard to the Canadian public, appropriate action, including a recommendation to deregister (the chemical), will be immediately initiate."

Just two weeks before that, Begin had given assurances there was no evidence of that potential health hazards existed with any of the IBT chemicals.

* * *

Laws Covering Chemical Labelling Said Poor

Industrial chemicals can be killers.

Some victims die quickly. Others go slowly with one of a variety of agonizing diseases. Some victims live, but spend life as cripples.

Yet Canadian labor laws provide little protection for workers exposed to the thousands of chemicals used in the workplace, Bob Sass, director of occupational health and safety with the Saskatchewan labor department, said.

Yet because of poor laws covering the labelling of chemicals, most workers do not have the information they need to protect themselves. He said many chemicals in use in industry are known to be dangerous, but the use of trade names for products containing these chemicals often hides their true nature from workers, employers and government officials.

"There are numerous deficiencies in labelling of chemicals in Canada," Sass said.

"Many of the chemicals that go into the workplace cannot be innocuous. They have been brought in to do something – they are reactive, they are agents that make changes and do things. That's why they are used.

"I'm not saying they are all equally harmful. But if you look at the history of occupational health, more and more is becoming known about their toxicity and their long-term hazards."

The most dramatic recent examples of chemicals killing or injuring workers are Kepone, which attacks the central nervous system of workers exposed to it, and Vinyl Chloride, a powerful cancer-causing chemical. The effects of Kepone were such that its use has been banned in Canada and the United States. After the hazards of Vinyl Chloride were discovered, the permitted levels of the chemical in air in plants using it were severely cut.

More chemicals now in use will be found to be dangerous, Sass predicted,

Moves towards tougher labelling requirements are being made in the United States. The Canadian government should at least keep pace, if not lead the way, or it could become a dumping ground for the hazardous chemicals not permitted in the U.S., he said.

The giant multinationals are not above shipping problems to areas with less stringent regulations. A number of companies have moved asbestos production facilities to countries where regulations are more relaxed than in North America, exporting occupational disease at the same time, Sass said,

"I don't think this is so crazy. There are many of these measures in consumer protection. It seems to me we ought to go at least that far to protect workers, and then a bit further, because we know what kind of stuff they are handling."

He said Saskatchewan is moving towards the introduction of laws requiring all employers to list the chemicals they use, the known health hazards of those chemicals, and a description of what is not known about the health hazards. All that information will have to be made readily available to workers.

But provincial action on labelling of containers of chemicals is nearly impossible, Sass said.

Jim McLellan, director of occupational health and safety with the federal labor department, said the Canadian dangerous substances regulations do provide the type of protection for workers that Sass is advocating.

But those regulations only apply to workers in occupations under the jurisdiction of the federal government – about eight per cent of the Canadian workforce. McLellan said the federal government, in cooperation with the provinces, is moving towards legislation that will better protect the remaining 92 per cent of the work force. He did not have any idea of when that protection would be in effect, however.

Sass notes labelling of pesticides is now much more stringent than for industrial chemicals.

But there are still serious inadequacies in that area as well. Dr. David Penman, senior health consultant with the Saskatchewan environment department, said.

He said warnings that people should not breathe or have skin contact with pesticides are not sufficient.

"The labels should fully advise people of the potential health effects of exposure to these chemicals."

Many of the pesticides in use can be absorbed through the skin, yet labels usually give no indication of that.

Major differences in labelling of identical pesticides exist between Canada and the U.S. The herbicide TOK banned by Saskatchewan because it causes cancer, birth defects, and other health problems, is one example.

On the drums available in Canada the warning reads: "Don't breathe spray mist. Avoid swallowing and contact with skin and eyes." There is no special warning to women, and no indication of possible health hazards.

The label Oil drums of the chemical sold in the U.S. reads: "Restricted use pesticide. For retail sale to and use only by certified applicators or persons under their direct supervision.

"WARNING TO FEMALE WORKERS. Women of childbearing age should not be involved with mixing/loading or application of this product. Exposure to this product during pregnancy must be avoided. Exposure to TOK has caused severe, delayed adverse effects in experimental animals.

"Severely irritating to the skin and eyes. Readily absorbed through the skin. Harmful if swallowed, inhaled, or deposited on skin."

The warning goes on to say special protective clothing must be worn when using the chemical and anyone unwilling or unable to follow all the warnings on the label must not handle the chemical.

All labels on pesticides used in Canada must be approved by the federal agriculture department.

But comments made by an official of that department at the annual meeting of the Canadian Association of Pest Control Officers, an organization of federal and provincial officials, last November raise serious questions about the attitude of the federal government.

In a document obtained by The Leader-Post, Wayne Ormrod, chief of the federal agriculture department's pesticide section, said product labelling "should not get too detailed in light of the need to treat large and small areas" of forested land.

He went on to say too much detail on labels could lead to increased violations of the law or to more charges from environmental groups. Labelling will be handled in such a manner as to avoid such problems, he said.

At the same meeting it was noted labelling problems with drums of the pesticide picloram (tradenamed Tordon 22K) resulted in the chemical being sprayed from aircraft near Penticton, B.C. Drift of the chemical caused damage to between 80 and 100 properties.

B.C. officials at the meeting said the chemical might be safe for aerial application in wide open areas, but it was not safe for application in that manner near settled areas. Yet the labels on the chemical containers gave no indication the chemical should not be applied using aircraft.

Published: The Regina Leader-Post, June – November, 1980

* * *

Safety Last: Tests that Fail the Test

It was the kind of story that elicits gasps. A woman in Oakville, Ont. was doing her supermarket shopping one day in mid-August when she noticed an employee calmly spraying the fruit with an insecticide. Horrified, she lodged a formal complaint. While the incident was quickly dismissed as harmless and "just plain dumb," it held special interest in light of recent reports that pyrethrin, a chemical in the spray, is one of more than 100 pesticides still on the Canadian market although their safety tests have been deemed invalid. Indeed, Canadians are getting doses of all sorts of chemicals with no assurances of safety. And furthermore, Canadian officials have known about the dangers for years but have chosen to do little and say even less.

Pesticides are approved for sale in Canada on a sort of honor system. The chemical company that develops a product is responsible for paying an independent lab to test for short- and long-term health hazards. The results are then submitted to the federal government for registration in Canada. Short of going to the considerable expense of duplicating all testing, the government has little choice but to trust that the procedures have been appropriate and the results honest. And this is where the honor system has broken down. In the past decade alone, separate cases of fraud involving more than a dozen labs or chemical companies have been uncovered in the United States.

The most serious recent case involves more than 100 pesticides tested by the American testing facility Industrial Biotest Laboratories (IBT). Washington's Environmental Protection Agency (EPA) is investigating suspected fraud and incompetence in the procedures employed by the company to test possible long-term health hazards. As well, EPA spokesman Jim Sibbison says the agency is auditing test results from "nine or 10 laboratories because of the possibility of careless or deliberately careless work."

Although problems with IBT'S testing were first discovered in 1977, Canadian officials didn't release the list of suspect chemicals until mid-June – after three months of repeated inquiries from Saskatchewan's environment department, news media and MPs. Included on the list (which is still growing) are: dichlorvos, used in "no-pest" strips and some flea collars; diazinon, a popular home and garden insecticide; atrazine, widely used on corn crops in Ontario; and fenitrothion, the controversial spray used in New Brunswick's spruce budworm control program. Dr. David Penman, a health officer with the Saskatchewan environment department, says that by mid-August his department still had not received satisfactory information about how IBT tests were flawed. "When people ask us what is wrong with these chemicals, the answer is, ridiculously enough, 'We don't know.'" The Ontario government has taken a different stance. "We have known all about this for the past three years," says George Cooper, chairman of the province's pesticide advisory committee. "But we feel it is a federal matter."

Dr. Alex Morrison, head of the federal health protection branch, is not overly concerned. "If IBT is an aberration we can deal with it," he says. "But a widespread inability to trust them (laboratories and chemical companies) could mean the government stepping in." He hasn't elaborated on what he means by "stepping in," even though the government has already spent $600,000 investigating IBT alone. But that figure pales beside the $100 million Morrison estimates it will cost – the chemical industry to retest the suspect pesticides discovered so far. He says that pulling all the chemicals off the market until the retesting is completed four years from now would have severe economic effects on Canadian agriculture as well.

The problem has prompted new demands that the federal government establish its own testing facilities. Regina East MP Simon de Jong, the NDP's science critic, says private laboratories have proven they are simply not morally capable of doing the job. He is critical of a recent statement by federal Health Minister Monique Begin that Canada is a small country that would have trouble testing all the chemicals in use here and should therefore take advantage of whatever information it can get elsewhere. Says de Jong: "The information we have been obtaining from other countries has proven to be false. The argument in favor of our own independent, trustworthy labs is just plain common sense."

Published: Macleans, August 25, 1980

* * *

Chemical Safety Tests: The Doubts Remain

Industry and the EPA have papered over evidence of dioxin contamination from pulp and paper mills.

The classic argument of the proponents of pesticides is that the chemicals have been extensively tested and found safe. But most of the pesticides sprayed in Canada were tested for safety under a system riddled with problems. Methods of testing the safety of pesticides in the United States are now known to be questionable and careless at best, fraudulent at worst.

Documents obtained from the U.S. Environmental Protection Agency (EPA) and the U.S. Food and Drug Administration (FDA) reveal serious shortcomings in the work at nearly 30 laboratories. The problems include:

- • falsification of research reports
- • failure to report tumors and other adverse effects
- • record-keeping errors that made scientific studies useless
- • improper administration of chemicals being tested
- • studies deliberately designed to minimize adverse health effects in test animals

There are problems with the work of an additional 22 laboratories, the EPA says.

To date, only details on scandalous test fixing by Industrial Biotest Laboratories (IBT) have been made public by the Canadian and American governments. For more than a year, officials of the chemical industry. and the Canadian government – which is heavily dependent on American testing – have been insisting that the IBT scandal was an aberration.

However, studies on the safety of more than 100 pesticides have been cast into doubt by the IBT scandal. Precise figures on the number of chemicals involved continue to fluctuate as government investigators dig into hundreds of thousands of pages of records. While a number of chemicals have been "cleared" by the Canadian government, the validity of tests on many of the most important pesticides used in Canada – including Roundup, picloram, diazinon, and toxaphene disulfaton – remains in doubt.

American officials say that changes in their testing system in 1977 have greatly improved the way things are done. But Ernest Brisson, director of the FDA's biomonitoring staff, acknowledges that most chemicals that are now on the market were test under a system that had many problems.

Even the new American system will not prevent companies from fabricating scientific studies and getting away with it, says a senior EPA official, who asked that he not be identified because of the Reagan administration's order that civil servants not talk to the news media.

Critics of the system maintain that the testing business should be taken out of the hands of the chemical industry. New Democrat MP Simon de Jong suggests that the Canadian government should establish its own testing facilities and work through the United Nations. The proposal would have the chemical industry pay the full cost of testing, as it does now, while removing the opportunity for fraud and manipulation of results.

Published: Harrowsmith, August-September 1982

* * *

Whitewash: The Dioxin Cover-Up

In the dying days of the summer of 1987 a ramshackle house in an isolated valley in Oregon's Coastal Range became the focus of international media attention. It was long in coming. Since the mid-1970s, Carol Van Strum had almost single-handedly conducted a vigorous campaign to end the use of dioxin-contaminated herbicides in her home state. The decade of struggle, chronicled in A Bitter Fog, might well have ended with the completion of the book. Except that Van Strum and her husband of eight years, attorney Paul Merrell, soon realized that the herbicide story was just part of something bigger.

Information on dioxin, a chemical that can be as dangerous as plutonium, was scattered and incomplete, they discovered. Even worse, EPA officials were reluctant to release publicly their own studies, some of which showed a connection between dioxin and paper mills. Using freedom of information laws, Van Strum and Merrell obtained EPA's dioxin studies in 1986 and set to work piecing together a report that pulled the dioxin/paper mill connection from behind, in the authors' words, "a smokescreen of government secrecy."

In August 1987, the report, entitled No Margin of Safety and published by Greenpeace, burst like a bomb on the pulp and paper industry and its regulators within the Environmental Protection Agency (EPA). Evidence gleaned from thousands of pages of the EPA's own documents demonstrated that pulp mills were spewing dioxins into the air and water, creating what Van Strum and Merrell call a public health emergency.

But that was only the beginning. Someone inside the American Paper Institute (API), the paper manufacturer's trade organization, saw the report and sent a collection of documents to Greenpeace. These documents substantiated Merrell and Van Strum's charges that senior EPA officials and the industries the agency was supposed to regulate were working together to limit public knowledge about the hazards of dioxin and a host of other dangerous chemicals. According to U.S. District Judge Owen M. Panner, the documents revealed an agreement "between the EPA and the industry to suppress, modify or delay the results of the joint EPA/industry [dioxin] study or the manner in which they are publicly presented."

* * *

Since at least 1980, EPA scientists and researchers with Canada's environment and health departments have been expressing their concern about the growing dioxin contamination of the environment. They are concerned about the high toxicity of dioxin and its extreme ability to bioaccumulate. Dioxin is the term commonly used to describe a group of about 75 compounds with the same basic chemical structure.

2,3,7,8-tetrachlorodibenzo-p-dioxin (TCDD) is the most studied member of the dioxin family. TCDD is also the deadliest substance ever produced. Its toxicity has been compared to plutonium-the EPA's procedures for handling these two materials are the same.

Industry representatives have argued that low levels of dioxin do no harm, but this contention has never been supported by scientific research. During congressional hearings in 1980, EPA scientists testified that TCDD was so powerful a carcinogen and teratogen that even the lowest measurable doses caused cancer and birth defects during laboratory tests. "EPA considers dioxin a carcinogen and, as all carcinogens, considers there to be a finite risk at any level," said the EPA's Alec McBride. "EPA considers any level as posing a degree of risk."

Cancer is not the only danger that dioxin poses. The effects of TCDD in all species of animals tested under laboratory conditions included weight loss, liver damage, hair loss, abnormal retention of body fluids and suppression of the immune system. Other effects of exposure to TCDD include birth defects and infertility. The dangers to the unborn in particular were emphasized by the EPA's Don Barnes. In a memo written on March 16, 1987, he said:

"Pregnant women, lactating mothers, developing fetuses and nursing infants constitute a subpopulation of special concern. Human body burdens [of dioxins and furans, a closely related group of highly toxic chemicals that are often found with dioxins] are likely to lead to additional burdens to the fetus and the nursing infant, which are not mimicked in the animal tests. Increases in the mother's body burden as the result of [dioxin/furan] contaminated food would likely lead to additional exposures."

In the late 1970s and early '80s, public concern over dioxin contamination centered on sites like Love Canal and Times Beach. But it soon became evident that dioxins are far more widespread in the environment than two places in New York and Missouri.

* * *

In the forests of Van Strum and Merrell's valley, and in many others up and down the Coastal Range, the spraying of defoliants similar to Agent Orange was commonplace during the 1970s and into the early 1980s. So were health problems that many people complained were the result of that spraying.

"One woman had had fourteen miscarriages in the years she had lived in the valley. Another told of her two miscarriages and of her son born with defective lungs and liver. The young wife of a logger had been unable to complete a pregnancy in the five years they had been married," wrote Van Strum in her book.

In Oregon, Larry Archer and his wife Laura lived near a reservoir that was sprayed with herbicides while she was pregnant with her second child. He was present when the baby was born. Van Strum wrote, " 'The baby – it was a girl,' Archer said. 'She was perfect, from her toes to her eyebrows. I mean, her face was perfect too... But that was it. It ended at the eyebrows. That's all there was – just this kind of a bowl, with a kind of tissue over it. She couldn't breathe. There wasn't any brain to tell her to breathe."

Van Strum, Merrell and others say the evidence clearly shows a link between herbicide spraying programs, dioxins and birth defects. What concerns them is that dioxins have been detected in places where little or no spraying of herbicides has been done and from a variety of sources. Dioxins are showing up everywhere in the environment – and in the food chain.

Dioxins are always produced when chlorinated compounds are burnt. Municipal incinerators, for example, produce dioxins when they burn garbage containing chlorinated plastics like polyvinyl chloride (PVC). Dioxins are also unwanted by-products in the manufacture of chlorinated chemicals, such as Agent Orange and the wood preservative pentachlorophenol (PCP).

EPA scientists suspected some sort of link between pulp mills and dioxin in 1980. Their suspicion was substantiated in 1983, when fish caught downstream from several Wisconsin River pulp mills were found to contain high (50 parts per trillion) levels of dioxin. The dioxin studies secured by

Van Strum and Merrell in 1986 confirmed the cause-and-effect relationship. Samples from sites slated for "control" sampling and predicted to have only "background" dioxin levels consistently revealed high levels of dioxin contamination when downstream from or near pulp mills.

Chlorine in the pulp bleaching process acts to form toxic chlorine-based compounds, including dioxin. Kraft-type pulp mills, where chlorine gas is used in the first stage of the bleaching process, are the biggest culprits. Chlorine gas reacts with compounds in wood lignin to create dioxin precursors and many other chlorinated compounds. These toxic compounds, called organochlorines, are released into the air and water when wastes are dumped. In North America, more than 150 pulp mills are dumping organochlorines, and most likely dioxins, into nearby rivers and lakes. An average-sized pulp mill discharges between 35 and 50 tons of chlorinated compounds every day.

"Dioxin, really, is only the tip of the iceberg when it comes to pulp mill effluents," says Renate Kroesa, Greenpeace's international pulp and paper campaign director. "Up to 1,000 different chlorinated compounds, only 300 of which have been identified, are formed during bleaching and discharged with the effluent. Among the compounds identified, we find many well-known carcinogens that are regulated when they come from chemical industries. But, when they are discharged by the pulp and paper industry, there are no limits."

* * *

The island scattered in the Georgia Strait between the British Columbian mainland and Vancouver Island are close to heaven on Earth. Sparkling blue water, the deep green of fir, spruce and hemlock and rocky shores make it a beautiful area. The waters around the islands teem with life. Pods of orcas are not an uncommon sight. Crabs, oysters and shellfish abound. But the beautiful scenery is marred by foul air at Crofton, Vancouver Island, where a pulp mill vents its wastes into the air and water.

Like the canaries that once warned coal miners of deadly gases in their pits, the Crofton blue heron colony serves as a warning of new dangers. For the last two years the eggs from the blue heron colony near Crofton have failed to hatch. Dioxins were detected in the herons' eggs during studies conducted in 1983 and 1986. The dioxins came from two sources: hexa-, hepta-, and octa-dibenzo-p-dioxins were coming from lumber industries in British Columbia that use dioxin-laden pentachorophenol wood preservatives; TCDD was coming from the pulp mill. Since PCP use has decreased since 1987, levels of dioxins associated with it have dropped. But the pulp mill operates beyond capacity, so levels of TCDD dioxin have tripled over the last two years.

Crofton's paper mill is an example of an industry-wide problem. Dioxins not only find their way into the water. They also contaminate the bleached pulp produced by these mills. What this means is that all bleached paper products – coffee filters, disposable diapers, toilet paper, everything – are potentially contaminated with dioxin. Paper industry executives would prefer to keep this quiet. And they almost did, were it not for the ally who sent the American Paper Institute's internal papers to Greenpeace.

The API papers show a concerted effort on the part of the EPA and industry to hide the problem. The strategy began with an effort to slow the introduction of regulations aimed at eliminating dioxin contamination. In an internal memorandum to the American Paper Institute's executive on December 30, 1986, one staff member claimed that "the industry has been able to forestall major regulatory and public relations difficulties by, among other things, agreeing to cooperate in a joint study with EPA."

In addition to forestalling regulatory action, the paper industry put together a "dioxin response team" which recommended "a public affairs strategy calling for activities keyed to, and in advance of, the release of the joint EPA/industry dioxin study."

A comprehensive API plan dated March 2, 1987, treated the public health threat posed by dioxin as a public relations problem. The industry's strategy was to: "1) keep all allegations of health risks out of the public arena – or minimize them; 2) avoid confrontations with government agencies, which might trigger concerns about health risks or raise visibility of issue generally; 3) maintain customer confidence in integrity of product; and 4) achieve an appropriate regulatory climate."

In the same document the industry said it would meet with EPA administrator Lee Thomas to tell him that dioxins pose no real health hazard and that the problem is one of "public perception." One of the industry's objectives was to have the EPA "rethink" its dioxin risk assessment and issue a statement saying dioxin causes "no harm to environment or public health."

In its efforts to sway the EPA, the American Paper Institute planned to turn EPA attention away from dioxin contamination of paper products. At the same time, it was working to "improve intelligence gathering within EPA" including identification of "allies" and "adversaries" within the agency.

According to an API report distributed to its members on August 10, 1987, the industry achieved at least some of its public relations goals: "[EPA] Administrator [Lee] Thomas indicated a willingness to cooperate with the industry to ensure that the public would not be unduly alarmed about this [dioxin] issue."

In its efforts to stifle public discussion of the dioxin hazards posed by its plants, the pulp and paper industry even attempted to get the EPA to violate the freedom of information laws. In a sworn statement given May 2, 1988, the EPA's McBride said the paper industry had specifically asked that the agency not release information to Van Strum, as it was required to do by law.

A letter written on March 18, 1987, by Carol Raulston, the API's vice president for government affairs, to a public relations firm employed by API, indicates that these efforts to interfere with freedom of information laws were successful. She wrote: "EPA has agreed to the following: to characterize the information (about dioxin contamination) as meaningless, used only to establish testing procedures; to respond with a letter to Ms. Van Strum today, but to ship the material on April 1; to meet with us to discuss the public affairs strategy on this and how subsequent requests for information will be handled."

When questioned about this, EPA's McBride jumped to the defense of his agency and said, "These were not agreements. When they say EPA has agreed to the following, these are EPA's decisions independent of anything we heard from the paper industry. Clearly what you have is a lobbying firm that has failed in its primary objective, which was to have us not release the data. They are trying to make it look as if they have accomplished something."

Although the pulp and paper industry was trying to stifle the release of government information, it discovered some disturbing new revelations about the dangers that dioxin contamination of paper products presented. TCDD, the most deadly of the dioxins, was found in bleached pulp at levels ranging from one part per trillion to 51 parts per trillion (ppt) in the vast majority of the samples taken. Levels of related chemicals were found to range from 1.2 ppt to 330 ppt.

Part of the industry's public relations strategy was to dismiss these levels – "trace amounts" as they have often been called – as being far below any level that presented danger to the public. In a speech to an API industry forum on March 8, 1988, Thomas C. Norris of P.H. Glatfelter Company, an API member, told his colleagues that the results of the industry's testing work into the dioxin problem had been "extremely encouraging."

"First, the dioxin detection levels are quite low," Mr. Norris said. "They range from no detection in the disposable diaper sample to 3.8 parts per trillion in paper towels to 14 parts per trillion in non-barrier food packaging." He went on to say that tests done for the industry had demonstrated only minimal movement of dioxins and related chemicals from paper products to the human body. "We are very encouraged by these results, and the bottom line is that all of the testing work done to date confirms that our paper products are safe," Norris said.

Yet other tests sponsored by the API itself showed that superabsorbent disposal diapers had up to 11 parts per trillion of dioxin in them, paper towels up to 7 parts per trillion, and various types of paper plates up to 10 parts per trillion. A draft report prepared for the industry in June 1987 by the research firm of A.D. Little found that between 50 and 90 percent of the dioxins in paper products "in contact with food oils or water is available for consumption."

Even more damning words about the hazards of dioxin in paper products were then being written by EPA officials. "If the exposure estimates utilized in the risk assessment are reasonably accurate-and I have no reason to believe they are not," wrote EPA scientist Dr. Fran Gostomski on July 10, 1987, "we are presented with a risk estimate for at least one exposure scenario-ingestion of dioxin from coffee filters-that exceeds the lifetime risk level at which

EPA would generally be expected to take regulatory action. In addition, this risk estimate does not take into account the very probable occurrence of simultaneous exposure to multiple sources of dioxin from bleached kraft paper products."

If one were to take milk or cream with that coffee, it would add even more dioxin to the diet. In the summer of 1988, at the International Dioxin Symposium, the Canadian Health Protection Branch of the federal health department presented evidence showing that dioxin in paper milk cartons had migrated into the milk. This was despite paper industry assurances that it was impossible for dioxins to migrate from cartons.

Although EPA is aware of the connection between pulp mills and dioxins, the agency has failed to produce regulations that would eliminate dioxin contamination of air, water and paper products. Instead, in April 1988, the EPA decided to do another, bigger study of all 104 pulp and paper mills that use chlorine.

"It's totally unnecessary," says Van Strum.

"It's the classic 'further study' in place of taking any regulatory action. They know there are hazardous levels of dioxins and furans in pulp and paper products. There's simply no need for further study."

She said that after the initial study of five mills revealed dioxin contamination, the EPA drafted a number of regulatory actions. These drafts have never been translated into effective measures. "The industry flacks within EPA prevailed and put off action until further studies were done," Van Strum said.

The Environmental Defense Fund and National Wildlife Federation sued EPA for its complacency. Settling out of court, EPA agreed to complete a risk assessment of the 104 mills by April 30, 1990. But even after that is done, EPA can: 1) refer the problem to another federal agency; 2) decide that dioxin from bleached wood pulp doesn't produce an unacceptable risk; or 3) take another year (until April 30, 1991) to propose regulations.

* * *

While North America studies dioxin, several European governments have decided to deal with the problem head on. Throughout Europe, the need for highly bleached paper products is being re-evaluated. Sweden, for example, has stopped the sale of chlorine-bleached disposable diapers. In Austria, consumers are using unbleached brown coffee filters and milk cartons.

"Household products are one of the most important keys in the struggle against environmentally unsound consumer goods," said Brigitta Dahl, Swedish minister of the environment. "Therefore we are now concentrating our efforts against chlorine and dioxins in the most common household products. This will be a strategically important contribution. If one gets paper bleached with chlorine out of consumer products, one also gets large amounts of chlorine out of the industrial stage, and the consumers don't have to live with an environmental threat on their breakfast tables, in their bathrooms and in large parts of their lives."

Sweden is making great strides in getting dioxins and all other toxic organochlorines out of its pulp and paper mills. Today the Swedish pulp industry discharges about 3.5 kilograms of organically bound chlorine per ton of pulp. But new laws require that mills reduce their discharge to 1.5 kgs/ton by 1992 and completely stop it by the year 2000. Swedish mills are using oxygen bleaching, among other things, to meet this goal.

The North American pulp and paper industry has used delaying tactics to avoid legal liability for medical problems that people may have suffered as a result of exposure to dioxin, Van Strum said. She says the EPA is hesitant to regulate dioxins for the same reason.

"There is no question they are trying to avoid regulating the most toxic known substance. There are two reasons for this," she said. "First, if the EPA were to say dioxins were hazardous, they [would] create a liability in cases like the Agent Orange litigation. Second, several laws the EPA administers specifically state only safe doses of pollutants can be released into the environment. Saying dioxins are hazardous at any level would seriously affect many industries and activities."

Industry recognizes dioxin-contaminated paper as a problem. Unfortunately, some solutions being offered are inadequate, because they eliminate or reduce dioxins in the final product, but continue releasing them into the environment. For example, Dow Chemical Corporation is developing ion-exchange resins that would remove organochlorines from pulp. But if this method were implemented, the dioxin-saturated resins would still have to be disposed of somewhere.

A major political fight may be required to stop the spread of deadly dioxins and other organochlorines. Regulations that include a time table for zero discharge of these toxics seem far away. EPA policy makers continue to stall; they attempted to reduce their obligation to clean up dioxin-contaminated sites by announcing last year their intent to increase by 16 times – the levels of exposure that will be deemed acceptable. Fortunately, EPA's Science Advisory Board agreed at their December 1988 meeting that "there is no firm scientific evidence" for the proposal.

Greenpeace is pushing for standards that will completely eliminate organochlorine discharges by 1993. This can be achieved by abandoning the use of chlorine in the bleaching process. "What is needed is a lot of local participation by people," Paul Merrell says. "That is the only way that the spread of dioxins into the environment from pulp mills will be halted quickly," he adds.

In addition to political pressure, the economic weight brought by changing consumer demand for bleached paper products may be needed to force government and industry in North America to deal with the problem. Greenpeace has asked for the immediate introduction of chlorine-free and/or unbleached paper products, as well as a higher recycling rate of paper products. The North American industry has resisted these demands. Coffee filter producers, for example, claim that they do not have enough unbleached pulp to produce unbleached filters. Yet at the same time the pulp industry is undergoing an enormous expansion program, all geared to producing more chlorine-bleached pulp. Unbleached pulp is cheaper and easier to manufacture.

For years whiter-than-white paper products have been associated with hygiene by consumers. Now they should be seen as a threat to health and the environment.

"Paper is a natural product, made of a potentially renewable resource," says Renate Kroesa. "How can we ever come to terms with living on this plan-

et if we don't even produce paper in an environmentally sound way?"
Published: Greenpeace, March-April 1989

Bibliography and Notes

The Murder of Melyce

1. Merrell v. Block. 14 ELR 20225, No. 81-6138-E, s.l.: 20 ERC 1620/(D.Or.), 4 18, 1983.

2. [Online] [Cited: June 11, 2014.] http://openjurist.org/747/f2d/1240.

3. Smith, Jeffrey. EPA Halts Most Use of Herbicide 2,4,5-T. Science. 1979, Vol. 203, 4385, pp. 1090-1091.

4. Strum, Carol Van. Personal communication with Donald Barnes, (Carol Van Strum, personal communication with Donald Barnes on August 8, 1983, in Carol Van Strum, Paul Merrell, et al v. Lee M. Thomas in his official capacity as Administrator of the United States Environmental Protection Agency, U.S.D.C. Oregon, Civil No. 84-6484, Affidavit of Carol Van Strum, May 14, 1986, pp. 5-6.

5. https://www.alecexposed.org/wiki/Right_to_Farm_Act_Exposed)

6. https://katu.com/news/local/oregons-right-to-farm-law-questioned-in-pesticide-dispute

7. https://nationalaglawcenter.org/state-compilations/right-to-farm/

8. http://www.earthtimes.org/green-blogs/green-opinions/alec-lets-corporations-craft-environmental-policy-29-Jul-12/

9. https://www.oregon.gov/ODA/shared/Documents/Publications/NaturalResources/RightToFarm.pdf

10. Spence, Gerry. With Justice for None: Destroying an American Myth. New York: Crown, 1989. 0812916964.

11. Christie, Agatha. Hallowe'en Party. New York: Harper, 1969.

A Bitter Fog

(No notes.)

Agents Orange, White, and Blue

1. For a summary of the history of phenoxy herbicides' development, see Peterson, Gale E. "The discovery and development of 2,4-D," Agricultural History 41:3, 1967, pp. 243—253; see also Davis, Donald E. "Herbicides in peace and war", BioScience 29:2, 1979, pp. 84, 91—94.

2. The use of 2,4-D on the Mall continued until 1980, when it was halted because of the potential health hazard. See Omang, Joanne. "Park Service to stop 2,4-D weed killer use on all its 325 parks", Washington Post, November 15,1980, p. A-2.

3. U.S. EPA 2,4-D Fact Sheet (commonly referred to as EPA's 2,4-D Position Document), July 24, 1981. See also Federal Register 43:17118—19, April 21, 1978, for recent figures on 2,4,5-T production.

4. Hawkes, Carl. "Planes release tree plantation." Journal of Forestry, May, 1953, pp. 345-348.

5. Historical Office, U.S. Army Chemical Corps. Summary of Major Events and Problems, Fiscal Year 1955. December 1, 1956, p. 60. See also Summary of Major Events and Problems, Fiscal Year 1959 (same authors), p. 106: Among "the objects of the revived anticrop program" was "to maintain close contact with the United States Department of Agriculture, universities and other agricultural institutions in order that information of military value may be properly correlated." The cited documents are among those declassified as a result of a Freedom of Information Act appeal to the Department of Defense by Canadian Reporter Peter von Stackelberg of the Edmonton (Alberta) Journal.

6. Wade, Nicholas. "Viets and vets fear herbicide health effects." Science 204:25, p. 817, May 25, 1970. See also Uhl, Michael, and Ensign, Tod. GI Guinea Pigs: How the Pentagon Exposed Our Troops to Dangers More Deadly Than War: Agent Orange and Atomic Radiation. Playboy Press, 1980, p. 111.

7. Unpublished staff report, U. S. Senate Subcommittee on Oversight and Investigations. September 25, 1980. Document on file with National Veterans Law Center, 4900 Massachusetts Avenue NW., Washington, D.C. 20016.

8. GI Guinea Pigs, note 6, op cit., p. 140.

9. Baughman, Robert. "TCDD and industrial accidents." In Whiteside, Thomas. The Pendulum and the Toxic Cloud: The Course of Dioxin Contamination. New Haven: Yale University Press, 1979, pp. 145-158. Baugh-

man's review of the scientific literature is probably the best publicly available summary of the history of TCDD industrial exposure cases.

10. See e.g., Meselson, Matthew, Westing, A.H., and Constable, J.D. Background material relevant to presentations at the 1970 annual meeting of the A.A.A.S., "Herbicide Assessment Commission Report to the American Association for the Advancement of Science:" December 1970, revised January 1971. In "Hearings before the Subcommittee to Investigate Problems Connected with Refugees and Escapees:" U.S. Senate Judiciary Committee, 92nd Congress, 1st Session, April 21, 1971. Part I: Vietnam. See also The Pendulum and the Toxic Cloud, note 7, op cit.; Shoecraft, Billee. Sue the Bastards! Phoenix: Franklin Press, 1971; Nielands, J.B., et al. Harvest of Death: Chemical Warfare in Vietnam and Cambodia. New York: The Free Press, 1972; Whiteside, Thomas. Defoliation: What Are Our Herbicides Doing to Us? New York: Ballantine Books, 1970.

11. Bionetics Research Laboratories, Inc. 1969. "Evaluation of the carcinogenic, teratogenic, and mutagenic activity of selected pesticides and industrial chemicals." Vol. III. "Evaluation of the teratogenic activity of selected pesticides and industrial chemicals in mice and rats." Submitted under contracts PH 43-64-57 and PH 43-67-735 with the National Cancer Institute. The raw data developed by Bionetics was further evaluated in Mrak, Emil, et at., Report of the Secretary's Commission on Pesticides and their Relationships to Environmental Health, U.S. Dept. of Health, Education and Welfare, December, 1969. The Mrak Commission placed 2,4-D in the classification to be "immediately restricted to prevent risk of human exposure" (p. 285). Subsequent studies by the National Institute of Environmental Health Sciences, USDA, the Canadian Food and Drug Directorate, and the Dow Chemical Company "showed that 2,4-D, its various formulations, and one breakdown product produced teratogenic effects in mice, rats, and hamsters. Some of the common abnormalities observed were cleft palate; clubfoot; no eyes; no lower jaw; distorted, missing, or extra ribs; unformed breastbone; brain mislocation; enlargement of the skull; and various hemorrhages." The Other Face of 2,4-D, A Citizens' Report. South Okanagan Environmental Coalition. Penticton, B.C., 1978, 1979. The latter work is a good overview of problems with 2,4-D.

12. U.S. Senate Hearings before the Subcommittee on Energy, Natural Resources, and the Environment, Commerce Committee, 1970. "Effects of 2,4,5-T on man and in the environment." 91st Cong., 2nd Sess., 7 and 15 April 1970. Serial 91-60.

13. The regulatory history which follows is drawn from Citizens Against Toxic Sprays, et al. v. Bergland, et al. (U.S.D.C. Oregon), Civil No. 76-438, Opinion of March 7, 1977, pp. 6-13; 428 F. Supp. 908,914-18 (1977).

14. Author interview with Dr. Theodore Sterling, December, 1980.

15. Although officially withdrawn from use in Vietnam, unofficially the use of herbicides in Vietnam was continued past 1970. Unpublished U.S. Senate Subcommittee staff report, note 7, op cit.

16. Federal Environmental Pesticide Control Act, Pub.L. 92—516, October 21, 1972, 7 U.S.C. §135.

17. For a good legislative history of early pesticide regulation, see Staff Report to the Subcommittee on Administrative Practice and Procedure, judiciary Committee, U.S. Senate. "The Environmental Protection Agency and the regulation of pesticides." December 1976.

18. Costle, Douglas M., Administrator, U.S.EPA Decision and emergency order suspending registrations for the forest, rights-of-way, and pasture uses of 2,4,5-Trichlorophenoxyacetic acid (2,4,5-T). February 28, 1979.

19. U.S.EPA FIFRA Consolidated Docket Numbers 415, et al. In re: The Dow Chemical Company, et al.

20. U.S. EPA 2,4-D Position Document, note 3, op cit.

21. Dow Chemical Company estimated that in 1964, the U.S. pesticide market "pie" was divided thus: herbicides 61.1%, insecticides 28.3%, fungicides 6.7%, soil fumigants 1.2%, plant growth regulators 1.7%, and defoliants and desiccants 1.0%. Statement of John Donalds, General Manager, Agricultural Products Dept., Dow Chemical, in "Extension of Federal Insecticide, Fungicide and Rodenticide Act Hearings." June 8, 9, 1977. U.S. Senate Subcommittee on Agricultural Research and General Legislation, Committee on Agriculture, Nutrition and Forestry.

22. Robert Charlton, public relations manager for Dow's agricultural products department, said, "There's a scientific principle at stake here...if [2,4,5-T] can be restricted...then what other important chemical tools might also be unnecessarily restricted?" In Epstein, Aaron, "Human guinea pigs: dioxin tested at Holmesburg." Philadelphia Enquirer, January 11, 1981.

23. This was the first emergency suspension of a pesticide registration because of human health impacts. Previous suspensions had been for environmental effects. See, e.g., suspension of coyote poisons involved in Wyoming v. Hathaway, 525 F.2d 66 (10th Cir., 1975).

24. The lengthy administrative hearing and judicial review processes prior to cancellation seem incompatible with the notion that the pesticide law was seriously intended to protect public health.

25. Tschirley, Fred, et al. Dispute Resolution Conference on 2,4,5-T. Arlington.

26. American Farm Bureau Federation, 1979.

27. Barnes, Earle B., Board Chairman, the Dow Chemical Company. November 1, 1979 letter to Rev. Robert E. Roos, Chancellor, Roman Catholic Diocese of Albany, N.Y. In Minneapolis (Minn.) Northern Sun News, March, 1980, p. 10.

28. Dow Chemical Co. v. Douglas M. Costle (U.S.D.C., E.D. Mich.), Civil No. 76—10087. Memorandum, Opinion and Order of April 4, 1978.

29. A summary of 2,4,5-T health effects can be found in U.S. EPA. "Rebuttable presumption against registration and continued registration of pesticides containing 2,4,5-T." Federal Register 43:17143, April 21, 1978. Also see Appendix A to this book.

30. FIFRA § 12(a)(2)(P), 7 U.S.C. § 136j, prohibits pesticide tests on human beings without their informed consent.

31. Epstein, Samuel. "Environment and teratogenesis." In Pathobiology of Development. The American Association of Pathobiologists and Bacteriologists, 1973, p. 105.

32. Rotbart, Dean. "Track of a toxin." The Wall Street Journal, December 8, 1981, pp. 125.

33. Senate Report 95-1188, Federal Pesticide Act of 1978 Conference Report, pp. 35-36, §12. Curiously, much of the offensive language of the 1978 Act was hidden in the conference committee report, rather than placed in the act itself; but the legislative history of an act is normally relied upon by the courts whenever there is controversy over the interpretation of language in a statute.

Inadmissible Evidence

1. Kirk, James, Vector Control Section, State of Oregon Health Division. October 9, 1974 memorandum to Tom Harrison, Plant Division, Oregon Department of Agriculture.

2. Riley, Dr. James A., Corvallis, Oregon. January 27, 1976 letter to David Eves, Esquire, Corvallis, Oregon.

3. Kirk memorandum, note I, op cit. 4.

4. Ibid.

5. Morse Laboratories, Inc., Sacramento, California. "Lab Report No. 95736, E.M. Longyear samples," August 21, 1975.

6. This memorandum was a significant document in subsequent legal action. (See Chapter 8).

7. Johnson, Edwin L., Deputy Assistant Administrator, U.S. EPA Office of Pesticide Programs. January 30, 1976 letter to Carrie Longyear.

8. T. J. Starker was among the four members of the first graduating class of the newly established Oregon State College (then Oregon Agricultural College) School of Forestry in 1910. He taught at OSU School of Forestry from 1922 to 1942. 50 Years of Forestry. Anniversary publication, Oregon State College (now University), 1956.

Early Warning

1. Shoecraft, Billee. Sue the Bastards! Phoenix: Franklin Press, 1971, p. 450. 2

2. Ibid., p. 60.

3. Kay, Jane. Article series: "Life fades and 'they don't think you notice.' " "11 years later, Globe victims still seek truth about spray" September 21, 22, 1980 respectively. Tucson, Arizona Daily Star.

4. Sue the Bastards!, note 1, op cit.,p. 18.

5. Information on the McKusick, McCray, and Steinke families, and on Jim Beavers and his friends is from Jane Kay, note 3, op cit.

6. Sue the Bastards!, note 1, op cit., from introduction by Dr. Frank Egler, p. xxv.

7. Except as otherwise noted, information and quotations from here to end of Chapter 4 are drawn from Billee Shoecraft's Sue the Bastards!, note 1, op cit., and from Billee's communications to the author before she died.

8. Galston, Arthur. "On the use and misuse of science." Yale Review, Spring, 1971, pp. 461—462.

9. Unpublished U.S. Senate Staff Report, Chapter I, note 7, op cit,

10. Kay, Jane, note 3, op cit.

11. Ibid.

12. Ibid.

13. Billee Shoecraft, personal communication to author. Ida Honoroff, publisher of Report to the Consumer, Sherman Oaks, California, confirms a similar communication from Shoecraft. According to Honoroff, Shoecraft surreptitiously made an audio recording of her examination by Dr. Charles Heine, a physician acting for Dow.

14. Kay, Jane, note 3, op cit,

15. "Dow Chemical settles lawsuits by 5 families over herbicide charges." The Wall Street Journal, March 4, 1981, p. 19.

Drifting In The Wind

1. Wilkinson, Jackie, interview with author.

2. Pezzi, Ugo, interview with author.

3. Anderson, jean, interview with author.

4. Green, Gisele, interview with author.

5. Elam, Daniel, interview with author.

6. Chart from information supplied by Gisele Green.

Not Only Plants

1. Shearer, Dr. Ruth, interview with author.

2. GI Guinea Pigs, Chapter 2, note 6, op cit., pp. 111—112. See also Unpublished U.S. Senate Staff Report, Chapter 2, note 7, op cit.

3. GI Guinea Pigs, Chapter 2, note 6, op cit., P: 140.

4. Ibid., p. 147.

5. Baughman, Robert, Chapter 2, note 9, op cit.

6. Barnes, Earle B., Chapter 2, note 26, op cit.

7. Baughman, Robert, Chapter 2, note 9, op cit.

8. Historical information on chloracne is largely derived from Baughman, Robert, ibid., and from research reports cited in that work.

9. Epstein, Aaron, Chapter 2, note 22, op cit.

10. See U.S. Air Force documents, Appendix A to this book. See also The Other Face of 2,4-D, Chapter 2, note 11, op cit., which contains a good summary and literature review of 2,4-D toxicity.

11. Shearer, Dr. Ruth, interview with author.

12. Mrak Report, Chapter 2, note II, op cit, (Also on this list was carbaryl [Sevin] insecticide, now in vogue for extensive aerial treatment of gypsy moths and spruce budworm. Carbaryl is still registered for home use.)

13. See, e.g., Costle, Douglas M. (emergency suspension order, Chapter 2, note 18, op cit., p. 6) gives a figure of 9.3 million pounds of 2,4,5-T used annually. U.S.EPA 2,4,-D Position Document (Chapter 2, note 3, op cit., p.1) places annual 2,4-D distribution at more than 70 million pounds.

14. Encyclopaedia Brittanica, 1963, vol. 20, p. 318.

15. Ibid., vol. 14, pp. 202—203.

16. Ibid., vol. 7, P: 615. See also Baughman, Robert, Chapter 2, note 9, op cit,

17. Ashton, Floyd M., and Crafts, Alden S., Mode of Action of Herbicides. New York: Wiley—Interscience, 1973, pp. 276—284.

18. Ibid.

19. Ibid.

20. Woolley, D. W. "Probable evolutionary relationship of serotonin and indoleacetic acid, and some practical consequences therefrom." Nature 180, pp. 630—633, 1957.

21. Peterson, Gale E., Chapter 2, note I, op cit., pp. 251—52.

22. Epstein, Samuel S., Chapter 2, note 30, op cit.

23. The link between carcinogenicity, mutagenicity, and teratogenicity is discussed in "Inter—Agency Regulatory Liaison Group work group report on the scientific bases for identification of potential carcinogens and estimation of risks;' Federal Register 44:39869, July 6, 1979. This report, a good primer on the state of knowledge on cancer, was adopted as part of national cancer policy by the Carter Administration on September 28, 1979, but was never adopted by E.P.A.'s Office of Pesticide Programs.

24. Inter-Agency Regulatory Liaison Group Report, ibid., p. 39876: "The self-replicating nature of cancer, the multiplicity of causative factors to which individuals can be exposed, the additive and possibly synergistic combination of effects, and the wide range of individual susceptibilities work together in making it currently unreliable to predict a threshold below which human population exposure to a carcinogen has no effect on cancer risk."

25. See U.S. Air Force documents, Appendix A. See also Baughman, Robert, Chapter 2, note 9, op cit.

26. See Sjöden, P.O. and Söoderberg, U., "Sex-dependent effects of prenatal 2,4,5-Trichlorophenoxyacetic acid on rats' open—field behavior." Physiology and Behavior9(3): 357—360,1972; Sjöden, P.O., and Söderberg, U., "Long lasting effects of prenatal 2,4,5-Trichlorophenoxyacetic acid on open—field behavior in rats: Pre- and post-natal mediation." Physiological Psychology 3(2): 175—178, 1975; Sjöden, P.O., and Sjöderberg, U., "Phenoxy acetic acids: sublethal effects." In Chlorinated phenoxyacetic acids and their dioxins. Ecological Bulletin 27:149—164, 1978; Sanderson, C.A., and Rogers, L.j., "2,4,5-Trichlorophenoxyacetic acid causes behavioral effects in chickens at environmentally relevant doses." Science 211:593—95, February 6, 1981.

27. Woolley, D.W., note 20, op cit,

28. Sjöden and Söderberg (1978), note 26, op cit., pp. 161—162: "...confirms...that 2,4,5-T...induces widespread changes in membrane function that are particularly strong and long-lasting in the central nervous system. It also influences thyroid activity...When a single dose of 2,4,5-T is given to pregnant rats, the offspring show reduced serotonin levels when adult."

29. For a more in—depth discussion, see, e.g.: Sjöden and Söderberg (1978), ibid.; White, A., Handler, P., and Smith, E. L., Principles of Biochemistry. New York: McGraw—Hili, 4th ed., 1968, pp. 589—90; Woolley, D.W., note 20, op. cit.

30. Sjöden and Söderberg (1978), note 26, op cit., p. 161.

31. Robson, J. M., and Sullivan, F. M., "Analysis of actions of 5-hydroxytryptamine [serotonin] in pregnancy." Journal of Physiology 184:717—32, 1966. See also Poulson, E., et al., "Effect of 5—hydroxytryptamine and iproniazid on pregnancy." Science 131:1101—02, 1960.

32. Sicuteri, F., et al., "Sex, migraine and serotonin interrelationships:" In Clinical Pharmacology of Serotonin: Monographs in Neural Sciences 3:94—101. New York: S. Karger, 1976. [Especially pp. 94—96.] See also "Relief from migraine." Newsweek, September 10, 1979, p. 69

33. Peet, N., et al., "Plasma total and free tryptophan concentration in 'neurotic' and 'psychotic' depressive patients:' In Clinical Pharmacology of Serotonin, ibid., 3:33—36.

34. Coleman, M., ed., Serotonin in Down's Syndrome. New York: American Elsevier, 1973, pp. 149, 161, 165.

35. Sjöden and Söderberg (1978), note 26, op cit., p. 150.

Critical Mass

1. Uhrhammer, Jerry. Eugene (Oregon) Register-Guard, series of articles in 1973 and 1974: "Will publicity doom Agent Orange use?" June 3, 1973; "Agent Orange draws U.S. ban," June 5, 1973; "Outlawed chemical used on Oregon lands," June 7, 1973; "Five states involved in Agent Orange tests," June 8, 1973; "Ban on 'Orange' not absolute," June 11, 1973; "EPA studies data on 'Agent Orange,'" June 12, 1973; "EPA closes probe on chemical," June 15,1973; "Plan drawn up for destroying Agent Orange," May 11, 1974.

2. Newport (Oregon) News-Times, December 25, 1975.

3. Micropthalmia and congenital blindness occur in test animals exposed to 2,4-D. See, e.g., Bionetics and Mrak reports, Chapter 2, note II, op cit.; Sue the Bastards!, Chapter 4, note I, op cit.; and Kay, Jane, Chapter 4, note 3, op cit.

4. Doctor, Ken. "Herbicide proponents reply; case goes to U.S. Court." Willamette Valley Observer, June 11—24, 1976, p. 4. The law firm was Davies, Biggs, Strayer, Stoel and Boley of Portland, Oregon.

"Too Many Unanswered Questions"

1. Baughman, Robert, and Meselson, Matthew, "Analytical method for detecting TCDD (dioxin): Levels of TCDD in samples from Viet Nam" Environmental Health Perspectives 5:27-35, 1973.

2. O'Keefe, Patrick W. et al. "A neutral clean—up procedure for 2,3,7,8 tetrachlorodibenzo—p—dioxin residues in bovine fat and milk," Journal of the Association of Official Analytical Chemists 61(3):621—26, 1978.

3. Baughman and Meselson, note 1, op cit.

4. Costle, Chapter 2, note 18, op cit., pp. 26—27. The order noted in a footnote (p, 27) that Dow had obtained an injunction to prevent public disclosure of the raw data on this and other studies.

5. Davis, Chandler, "Poison accumulation revealed;' Salem (Oregon) Capital Journal, June 30, 1976.

6. O'Keefe, Patrick. Testimony in CATS v. Bergland (U.S.D.C., Oregon), Civil No. 76-438, transcript of proceedings, June 23, 1976, p. 36.

7. Exhibit 1 (Tables 13 and 14) to Patrick O'Keefe testimony, ibid., p. 36.

8. Ross, Ralph. August 5, 1975, U.S.EPA memorandum to Edwin L. Johnson, Deputy Assistant Administrator for Pesticide Programs: "Interim results of 2,4,5-T/TCDD residues." A copy of this memo had been sent to the Longyears by Johnson.

9. Van Strum, Carol and Stevens, letter, reprinted in 1976—77 U.S.D.A.

10. Forest Service Pacific Northwest Region Vegetation Management with Herbicides Final Environmental Statement, pp. 05—08.

11. Vegetation Management with Herbicides EIS, ibid., pp. 106—107.

12. O'Keefe, Patrick. Supplementary Testimony in CATS v. Bergland, note 6, op cit., August 9, 1976.

13. Hayes, John, "Mothers' milk yields a trace of dioxin;" Salem (Oregon) Statesman—Journal, P. 1-A, February 17, 1977; "Straub to take close look at controversial spraying," February 18, 1977, p. 1-A. See also "Hatfield speaks out against 2,4,5-T spraying." Statesman-Journal, March I, 1977. In December, 1976, six months after finding TCDD in the Siuslaw mother's milk, Dr. O'Keefe tested a second sample from the same mother and found no detectable levels of TCOO remaining in her milk; Ruff, Lorraine, "New tests fail to find toxicity in milk sample," The (Portland) Oregonian, February 26, 1977, p. 4—B.

14. O'Keefe, Patrick, note 11, op cit.

15. CATS v. Bergland, note 6, op cit., redirect and rebuttal testimony of Dr. George Streisinger, pp. 2—3.

16. See Uhrhammer, Jerry, Chapter 7, note 1, op cit.

17. Council for Agricultural Science and Technology, The Phenoxy Herbicides: Report No. 39, 1975. For an interesting perspective on this "independent" organization, see January, 1979, BioScience: "CAST-Industry tie raises credibility concern."

18. CA TS v. Bergland, note 6, op cit., Opinion of March 7, 1977, p. 43.

19. Whiteside, Thomas, The Pendulum and the Toxic Cloud, Chapter 2, note 9, op cit,

20. See Lee, Brian, "Herbicide under suspicion in Australia." New Scientist, October 10, 1978; see also "Baby mystery inquiry is reopened," New Zealand Herald. March 4, 1977; CATS Newsletter, July, 1977, p. 5; and letter from Drs. William Sare and Paul Forbes. New Zealand Medical Journal, January, 1972.

21. Halling, H. "Suspected link between exposure to hexachlorophene and malformed infants." Paper read at the 1978 international conference on health effects of halogenated aromatic hydrocarbons, June 21—30, 1978, at New York Academy of Sciences. See also "New HCP harm," The (Portland) Oregonian, August 10, 1978, p. 2.

22. Pintarich, Paul. "Scientists term dioxin 'most toxic chemical," The (Portland) Oregonian, December 9, 1976. See also Whiteside, Thomas, The Pendulum and the Toxic Cloud, Chapter 2, note 9, op cit., pp. 133—34.

23. Tung, T., et al., "Le cancer primaire du foie au Viet Nam." Chirurgie 9:427-436, 1973. In Westing, Arthur H. Ecological considerations regarding massive environmental contamination with 2,3,7,8-tetrachlorodibenzo-para-dioxin. "Chlorinated Phenoxy Acids and Their Dioxins: Ramel, C., ed. Ecological Bulletin 27:285-294, 1978. See also Wade, Nicholas, Chapter 2, note 6, op cit.; and Smith, S., "Vietnamese doctor links spray, cancer," Eugene (Oregon) Register-Guard, April 28, 1979, p. 4A.

24. Kurtis, Bill, et al. March 23, 1978. "Agent Orange: Vietnam's Deadly Fog," WBBM-TV. Chicago, Ill. See also GI Guinea Pigs, Chapter 2, note 6, op cit.; and article by same authors in The Progressive, June, 1978. The story of Vietnam veterans and Agent Orange was initially broken by WBBM- TV, and quickly spread internationally.

25. See Chapter 11.

26. "U. S. Forest Service spray program shelved," Salem (Oregon) Statesman-Journal, March 9, 1977, p. 11A.

27. Hankins, Cathy. Personal communication, 1977.

28. CA TS v. Bergland, note 6, op cit., testimony of Mike Vernon Hankins. (Major portions of Hankins' testimony were reprinted in NCAP News, Fall, 1980, p. 38, under heading, "Where have you heard this before?")

A Soldier Turns to Home

1. Merrell, Paul, and Savage, Cheryl. Toolkit: Copy it! Edition I1A, April, 1979. Citizens for Alternatives to Toxic Herbicides (formerly Citizens Against Toxic Herbicides). Route 4, Box 2-A, Asotin, Washington 99402.

2. Operation Ranch Hand. 1966. "Special Aerial Spray Flight Handout": "If the temperature is above 80°F., the defoliant will not settle to the jungle target as planned."

3. Agent Orange Special Supplement to The Veteran, Spring, 1980. (Publication of Vietnam Veterans Against the War, Chicago, Illinois.) See also Unpublished U.S. Senate Staff Report, Chapter 2, note 7, op cit.

4. The "lunatic fringe" has an interesting history. A Forest Service employee, Roger Sandquist, was quoted in the March 3, 1978 Moscow (Idaho) Daily Idahonian as saying that most of the opposition to Forest Service use of herbicides was coming from "a lunatic fringe." Citizens Against Toxic Herbicides responded with a national fund-raising drive, selling membership certificates for "the Order of the Lunatic Fringe." See also "A Closer Look at Sandquist," NCAP News, Fall, 1980, p. 4.

"Those Chemicals from Vietnam"

1. GI Guinea Pigs, Chapter 2, note 6, op cit., in unnumbered illustration section.

2. All previous quotes in this chapter from: Agent Orange Special Supplement to The Veteran, Chapter 9, note 3, op cit. See also Bogen, Gilbert. "Symptoms in Vietnam veterans exposed to Agent Orange." Journal of the American Medical Association 242(22):2391, November 30, 1979: "A ten-month study of 78 Vietnam veterans who claimed exposure to Agent Orange yielded many findings. 85 of the men experienced a rash that was resistant to treatment....joint pain occurred in 71%, stiffness in 59%, and swelling in 45%. Hypersomnolence occurred in 44% of the men and extreme fatigue in 80% ... Persistent neurological complaints were tingling (55%), numbness (60%), dizziness (69%), headache (35%), and autonomic dyscontrol (18%). Severe psychiatric manifestations were depression (73%), suicidal attempts (8%), and violent rages (45%). An inability to concentrate occurred in 17%, and bouts of sudden lapses of memory...in 21% .... There was also a loss of libido in 47%. Three patients died of cancer. Another 10% have been treated for cancer."

Gastrointestinal problems in the study group included anorexia (41%), nausea (59%), vomiting (13%), hematemesis (8%), diarrhea (51%), constipation (31%), abdominal pain (24%). Hepatitis reported in 10%, jaundice in 5%. 19% fathered children with gross birth defects. 13% of wives had one or more miscarriages. Other frequent problems: blurred vision (54%), sterility, low sperm counts, abnormal sperm, frequent infections and allergies.

3. Weintraub, Pamela. "The Agent Orange mystery." Omni, August, 1982, pp. 14, 108.

4. Kurtis, Bill, el al., Chapter 8, note 23, op cit.

5. Uhl, Michael, and Ensign, Todd. "Blowing the whistle on Agent Orange." The Progressive, June, 1978.

6. GI Guinea Pigs, Chapter 2, note 6, op cit., pp. 194—95.

7. Kurtis, Bill, el al., Chapter 8, note 23, op cit,

8. U. S. EPA F.l. F. R. A. Consolidated Docket Numbers 415, et al., Chapter 2, note 19, op cit. January 13, 1981 direct testimony of Dr. Michael L. Gross. EPA Exhibit 930, pp. 12-17. (See chart, table 4, p. 14.)

9. Unpublished U. S. Senate staff report, Chapter 2, note 7, op cit, See pp. 1-5 of "History of Veterans' Actions Concerning Agent Orange" section.

10. Ibid.

11. Ibid.

12. Milford, Lewis M. "Justice is not a GI benefit." The Progressive, August, 1981, pp. 32-35.

13. Unpublished U.S. Senate staff report, Chapter 2, note 7, op cit.

14. Ibid.

15. Furst, Jon. Personal communication. The Task Force distributed health questionnaires to all Vietnam veterans contacted. Furst points out that because this is a self-selecting population, no firm conclusions can be drawn; but there is no apparent reason why men who worked on roadways in Vietnam should be more prone to health problems than other veterans.

16. Merrell, Paul. Author interview, 1980. The author has also reviewed files of correspondence between Merrell (on behalf of Citizens Against Toxic Herbicides), Senator Church, and the Department of Defense.

17. Code of Federal Regulations, Title 38, section 3.102.

18. Baughman and Meslson, Chapter 8, note 1, op cit.

19. GI Guinea Pigs, Chapter 2, note 6, op cit., pp. 169-171.

20. Baughman and Meslson, Chapter 8, note 1, op cit.

21. Ibid.

22. Ibid.

23. Texas protocols draft. Pilot study protocols for the Texas Veterans Agent Orange Program. December, 1981. Developed by faculty of the University of Texas for use in the Texas Veterans Agent Orange Program administered by the Texas Department of Health. 48 pp.

24. Unpublished U. S. Senate staff report, Chapter 2, note 7, op cit.

25. "Agent Orange legislation in Congress." Southern Coalition for the Environment Pesticide Newsletter 3(3):7, 1981. (Hammond, Louisiana.)

26. GI Guinea Pigs, Chapter 2, note 6, op cit., p. xiii.

27. Weintraub, Pamela, Chapter 10, note 3, op cit.

28. Suffer the Children, Chapter 12, note 4, op cit., pp. 194-205.

29. Unless otherwise attributed, information in the remainder of this chapter is based on publications of the Interfaith Center on Corporate Responsibility (I.C.C.R.) and of the National Council of Churches (N.C.C.), and on interviews with Bob Morris of I.C.C.R. and Rev. John Jordan of N.C.C. Both organizations are headquartered at 475 Riverside Drive, New York, N.Y. 10027.

30. Epstein, Samuel. The Politics of Cancer. San Francisco: Sierra Club Books, 1978.

31. Blustein, Paul. "Activist order: Maryknoll missionaries draw praise, criticism for social involvement." The Wall Street Journal, August 14, 1981, p. 1.

32. Weil, Simone. Waiting for God. New York: Putnam's Sons, 1951 Harper Colophon edition, 1973, pp. 98-99.

For Future Reference: The Alsea Record

1. Historical information on the Five Rivers Valley area is largely drawn from the folklore of "old-timers." Steve Tedrow, Beatrice Olson, and the late Roy Olson provided information especially useful in this chapter.

2. See, e.g., Elam, Daniel. Building Better Babies. Millbrae (California): Celestial Arts, 1980, pp. 79-96. See also Robson and Sullivan, Chapter 6, note 31, op cit.

3. Elam, Daniel, ibid. For example, Dr. Elam discusses research showing that thalidomide administered to male rabbits prior to mating accumulated in the sperm; offspring of subsequent matings were severely deformed. Page 89, with citation to Lutwak-Mann, C.; et al. "Thalidomide in rabbit semen." Nature 214:1018-20, 1967.

4. Mrak Report, Chapter 2, note 11, op cit. See also Galston, Arthur. "Herbicides: A mixed blessing." Bioscience 29(2):85-90, 1979.

5. Shearer, Dr. Ruth, interview with author.

6. The Maranos' story is based on the author's interviews with Debbie Marano, and on CATS Newsletter, December, 1977, p. 1.

7. The milk sample Debbie contributed was found to contain measurable levels of DDE, hexachlorobenzene, and Dieldrin. It was not analyzed for TCDD or phenoxy herbicide residues. See "Results from the national study to determine levels of chlorinated hydrocarbon insecticides in human milk, 1975—1976." Savage, Eldon P., Epidemiologic Pesticides Study Center, Colorado State University. Enclosure in Savage's September 21, 1978, letter to Debbie Marano.

8. Yoder, Julie, Watson, M., and Benson, W. W. "Lymphocyte chromosome analysis of agricultural workers during extensive occupational exposure to pesticides." Mutation Research 21:335-340, 1973.

9. See, e.g., figures given in USDA Forest Service, Pacific Northwest Region Final Environmental Statement, Vegetation Management with Herbicides, Vol. 1, 1978, pp. 434-35. The figures for the Siuslaw National Forest are a good example.

10. CATS v, Bergland, Exhibit 1 (Tables 13 and 14) to Patrick O'Keefe testimony. Chapter 8, notes 6 and 7, op cit.

11. Kearnev, P.C., et al. "Tetrachlorodibenzo-dioxin in the environment: Sources, fate, and decontamination." Environmental Health Perspectives 5:27 377, 1973. See also Helling, C.S., et al. "Chlorodioxins in pesticides, soils, and plants." Journal of Environmental Quality 2(2): 171-78, 1973.

12. See, e.g., In re: The Dow Chemical Company, et al. U.S.EPA FIFRA Consolidated Dockets 415, et al. Direct testimony of Dr. George Streisinger, EPA Exhibit No. 564, p. 33. Dr. Streisinger lists toxic effects such as still-births, alterations in thymus and kidney weights, and renal abnormalities.

13. Allen, James, et al. "Morphological changes in monkeys consuming a diet containing a low level of 2,3,7,8-tetrachlorodibenzo-p-dioxin." Food and Cosmetics Toxicology 15:401-410, 1977.

14. Jansson, Erik, interview with author.

15. Ibid.

16. Van Strum, Carol. "Ask the women." CoEvolution Quarterly, Summer, 1979, pp. 44-47. The quotation was from the author's prior interview with Dr. Kalishman.

17. Blum, Barbara, U.S.EPA Deputy Administrator, in press conference statement. "Emergency action to stop spraying of the herbicides 2,4,5T and silvex." U.S.EPA, March I, 1979, Washington, D.C., pp. 1—2.

18. O'Keefe, Patrick, Chapter 8, note 6, op cit.; Ross, Ralph, EPA memo., Chapter 8, note 8, op cit.

19. Hill, Bonnie, interview with author.

20. Sterling, Dr. Theodore, interview with author, December, 1980.

21. Newton, Jan. Testimony before U.S. House Agriculture Committee, Subcommittee on Forests, Family Farms, and Energy, field hearings at Eugene, Oregon, on problems and opportunities in forest management practices. January 3, 1980. Testimony reprinted in NCAP News, Winter/Spring, 1980, pp. 35-37. See also Final EIS, Siuslaw National Forest, Ten-Year Timber Resource Plan, USDA FS-R6-FES(Adm)-78-11, February 12, 1979, p. iii. Herbicide proponents acknowledge the connection between the allowable harvest and herbicide use in national forests: "The availability of the use of herbicides is assumed with each alternative" in setting allowable harvest levels. Put briefly, the allowable harvest level of timber is programmed ten years in advance on the assumption that herbicides will be used.

"The Siuslaw National Forest estimates that allowable cut would be reduced immediately by 34%...if herbicides were banned in that forest alone." (October 1980 campaign mailer, Lincoln County Citizens for Common Sense.)

Dr. Newton's research has been published by NCAP and is available in two parts: An Economic Analysis of Herbicide Use for Intensive Forest Management. Part I: Evaluation of Forestry Related Impacts of 2,4,5-T in Oregon. Part II: Critical Assessment of Arguments and Data Supporting Herbicide Use. NCAP, P.O. Box 375, Eugene. Oregon 97440.

22. The author finds it interesting that the Reagan Administration appointed Nolan E. Clark, a member of the

Washington, D.C., law firm of Kirkland & Ellis, a firm representing Dow Chemical in the 2,4,5-T cancellation hearings, to become EPA associate administrator for policy and resource management. See "Ann Gorsuch and the seven corporate lawyers." NCAP News, Spring-Summer, 1981, p. 5.

23. Rust, Gerald H., Jr. Testimony in April 16, 1981, Vegetation Management Reform Act of 1981 (H.R. 2900) Hearings, before U.S. House Committee on Agriculture, Subcommittee on Forests, Family Farms, and Energy. Field Hearings, Eugene, Oregon. The quoted statement did not appear in the hearing transcript, but was heard by those present, including the author, and was reported in the local press.

24. See, e.g., Epstein, Aaron, Chapter 2, note 22, op cit.

25. Merrell, Paul' "The low-down on Roundup." NCAP News, Fall, 1980, pp. 4-6. Merrell, Paul. "IBT officials indicted for fraud." NCAP News, Spring—Summer, 1981, P: 12. See also U.S. v. Joseph C. Calandra, et al. (U.S.D.C. N.D. Illinois, Eastern Div.) Special October 1980 grand jury indictment of June 22, 1981, p. I.

Poisons: Innocent Until Proven Guilty

1. Federal Environmental Pesticide Control Act (F.E.P.C.A.); Pub.L. 92-516, October 21, 1972.

2. See, e.g., U.S. Senate Subcommittee Staff Report, Chapter 2, note 17, op cit., for an excellent discussion of these problems.

3. Comptroller General. U.S. General Accounting Office Report CED-80-32, February 15, 1980. "Delays and unresolved issues plague new pesticide protection programs." pp. iv, 54-57.

4. The global scope of fraud involved in testing and marketing of thalidomide has been thoroughly examined in: Insight Team of The Sunday Times of London, Suffer the Children: The Story of Thalidomide. New York: The Viking Press. See especially pp. 79-85. Although thalidomide was never marketed in the U.S., 2,528,412 tablets were distributed to some 20,000 U.S. patients in clinical trials conducted by Richardson-Merrell, Inc. See Suffer the Children, p. 70. See also McCarthy, Frances. "Dow acquires ethical concern." NCAP News, Winter, 1981, pp. 6-7.

5. Lisook, Alan B., "FDA audit of investigators and sponsors." FDA paper presented at Drug Information Ass'n. workshop, December 9, 1981, Williamsburg, VA. The FDA document was obtained by Peter von Stackelberg utilizing the Freedom of Information Act.

6. Suffer the Children, note 4, op cit., pp. 64—68.

7. Rotbart, D., and Prestbo, John A. "Killing a product: Taking Rely off market cost Procter & Gamble a week of agonizing." The Wall Street Journal, November 3, 1980, p. 1. See also Rotbart, D. "State of alarm: Tampon industry is in throes of change after toxic shock." The Wall Street Journal, February 26, 1981, p. 1.

8. Suffer the Children, note 4, op cit., p. 37.

9. Unless otherwise attributed, information in this chapter is drawn from: von Stackelberg, Peter. "Fungicide not taken off market despite invalid tests." Regina (Saskatchewan) Leader-Post, November 4, 1980.

"Safety last: Tests that fail the test." Maclean's, August 25, 1980.

The Toxic Mist: the Use & Abuse of Pesticides. Collected articles. Regina (Saskatchewan) Leader-Post, 1980.

____ . (All from Edmonton [Alberta] Journal.) "U.S. problems with drug tests cloud our health," April 3, 1982; "U. S. agency critical of tests," April 3, 1982; "Warning on drug too late for many," April 4, 1982; "Shoddy work in U.S. drug tests," April 5, 1982.

Vonder Haar, T.A., et al. "The scandal at IBT." QUI, August, 1979, pp. 50-54, 122-123.

Merrell, Paul, Roundup article, Chapter 11, note 25, op cit.

____ . "The Industrial Bio—Test caper." NCAP News, Winter, 1981, pp.2-4.

Smith, R. Jeffrey. "Creative Penmanship in animal testing prompts FDA controls." Science 198:1227-29, December 23,1977.

_____ . "As luck would have it ..." Science 198:1228, December 23, 1977.

Doutt, Richard, "Debugging the pesticide law." Environment, December 1979, pp. 32-35.

Richards, Bill. "Probers say pesticide makers knew of faulty lab test data." Washington Post, March 9, 1978.

Lublin, Joanne, "A lab's troubles raise doubts about quality of drug tests in U.S." The Wall Street Journal, February 21,1978.

10. U. S. EPA press release. "Deficiencies in pesticide safety tests reported by EPA—Audit requested." August 25, 1977.

11. EPA has yet to release a complete list of pesticides and tests involved. A partial listing of approximately 200 pesticides obtained by Paul Merrell under the Freedom of Information Act is reprinted in NCAP News, Winter, 1981, p. 5, and also appears as Appendix D to this work.

12. Doutt, Richard, note 10, op cit.

13. For example, see EPA's discussion of "conditional" registration language proposed by the agency (later adopted by Congress) in "Extension of the Federal Insecticide, Fungicide, and Rodenticide Act;' Hearings before the U.S. Senate Subcommittee on Agricultural Research and General Legislation, Committee on Agriculture, Nutrition, and Forestry. 95th Cong., 1st Sess., June 8, 9, 1977, pp. 143-47. EPA officials made no mention of their knowledge of fraudulent testing in proposing that pesticides be registered without supporting safety data.

14. U.S.EPA "Industrial Bio—Test Laboratories, Inc. Audit Meeting." Transcript of EPA-Industry-Health Protection Branch Canada meeting at Howard Johnson Motor Inn, Arlington, Virginia, pp. 85, 88-89. This 130-page document obtained under discovery in Merrell v. J. R. Block is the most comprehensive view of IBT practices and the regulatory response to surface so far.

15. Union Carbide Agricultural Products Co., Inc., et al. v. Costle,—U.S.—, No. 80-1085, March 23,1981. Chevron Chemical Co. v. Costle,—U.S.—, No. 80-1680, June 22, 1981.

16. See, e.g., Merrell v. I R. Block, et al. (U. S. D.C. Oregon), Civil No. 816138-E, defendants' January 15, 1982, memorandum in opposition to motion to compel production. Also see Moore, John W, "EPA aims to halt release of pesticide information," Legal Times of Washington, February 22, 1982, p. 1. See also Smith, R. Jeffrey, "A battle over pesticide data." Science 217(4559):515-518. August 6, 1982. EPA's efforts to deny access to information came despite Deputy Administrator John Hernandez' July 22, 1981 testimony to Congress that the agency would begin making health and safety information available in September of 1981, a promise never fulfilled. FIFRA Hearing Report, U.S. House Committee on Agriculture, Subcommittee on Department Operations, Research and Foreign Agriculture, Part I, p. 605.

17. Merrell, Paul, Roundup article, Chapter 11, note 25, op cit.

18. von Stackelberg, Peter, interview with author.

19. "EPA to issue first data requests resulting from IBT test audit program." Chemical Regulation Reporter Current Reports, February 6, 1981, p. 1469.

20. Merrell, Paul, Roundup article, Chapter 11, note 25, op cit.

21. Merrell v. J. R. Block, et al. (U.S.D.C. Oregon) Civil No. 81-6138-E, Defendants' November 9, 1981, motion for summary judgment.

22. Ibid.

23. Shearer, Dr. Ruth, interview with author.

24. U.S. Senate Subcommittee Staff Report, Chapter 2, note 17, op cit., p. 43. EPA's misleading statements are apparently a continuing problem: In Merrell v. Block, a key government exhibit was an April 1981 "IBT Position Statement," signed by Edwin L. Johnson and representatives of the Forest Service and Bureau of Land Management. The document claimed that the IBT studies had been evaluated, as well as replacement studies. At a hearing on whether EPA would be forced to release the evaluations, Assistant U.S. Attorney Thomas C. Lee stated: "...he wants to have some assessment of all of the studies. Well, there isn't one .... EPA's evaluations are on-going." February 16, 1982, Transcript of proceedings, Merrell v. J R. Block, et al. (U.S.D.C. Oregon), Civil No. 81-6138-E, pp. 19-20.

25. Comptroller General 1980 Report, note 3, op cit., cover page.

26. von Stackelberg, Peter. "IBT not an aberration; fraud widespread." Regina (Saskatchewan) Leader-Post, June 6, 1981.

27. Johnson, Edwin L., U.S.EPA Deputy Assistant Administrator for Pesticide Programs. November 7, 1979, "Health effects data quality status report" memorandum to Peter McGrath, Director, Hazard Evaluation Division, with attached October 19, 1979, report, same title. (The documents were obtained by Peter von Stackelberg utilizing the Freedom of Information Act.)

28. "Nalco Chemical unit ex-officials charged with faking lab data." The Wall Street Journal, June 23, 1981, p. 21; Merrell, Paul, "IBT officials indicted:" Chapter 11, note 25, op cit.; U.S. v. Calandra, Chapter 11, note 25, op cit.

29. Could TCC be the mystery surfactant in herbicide Roundup? Vonder Haar, et al. (note 9, op cit.) reported FDA findings that TCC effects could include lesions of the testes, liver damage in offspring, brain damage, and injury to the spleen – effects mostly noted for the mystery surfactant. Merrell, Paul, Roundup article, Chapter 11, note 25, op cit. According to Vonder Haar and associates, TCC is found in deodorant soaps Zest, Palmolive Gold, Coast, Safeguard, Dial, and Irish Spring.

30. Health Protection Branch, Health and Welfare Canada, Ottawa. "Current recommendations on IBT pesticides:' Press packet, October 19, 1981.

31. Gross, Dr. M. Adrian. January 19, 1981, letter to Patricia Critchlow, U.S.EPA Office of Pesticide Programs Registration Division, with attached 73-page evaluation. The document is an example of what close scientific scrutiny can accomplish.

32. Johnson, Edwin L., note 27, op cit.

33. See U. S. Senate Subcommittee Staff Report, Chapter 2, note 17, op cit., p. 3 for estimates of increased use.

34. Doutt, Richard, note 10, op cit,

35. There is at least an appearance that the agency is not interested in existing knowledge of pesticide effects. As public interest in the IBT review program grew, Edwin L. Johnson decided: "We will not review an IBT study if the registrant has submitted, submits, or agrees to submit a suitable replacement study and he/she agrees to having the IBT study declared invalid for the purpose of future registration actions .... Studies which are considered invalid due to the lack of a sponsor validation, but have not yet been reviewed for possible adverse effects, will no longer be reviewed for adverse effects." Johnson, Edwin L., EPA Deputy Assistant Administrator for Pesticide Programs. April 29, 1982, form letter to pesticide registrants. (Document received under Freedom of Information Act by Paul Merrell.)

36. Johnson, Edwin L. (ibid.) Undated "Memorandum of concern regarding the regulatory follow-up of the IBT audit at Decatur, to A.E. Conroy, II, Director, Pesticides and Toxic Substances Enforcement Division, U.S.EPA The document was obtained under discovery in Merrell v. J. R. Block before the government moved for and was granted an order halting discovery until after summary judgment motions were decided.

The Five Rivers Health Study

1. Five Rivers residents. May 9, 1979 letter to Alsea Ranger District, Siuslaw National Forest, USDA Forest Service.

2. Two other women also miscarried during this time, but were not hospitalized and could not supply the "documentation" required to prove the miscarriages occurred.

3. Five Rivers residents. July 17, 1979, letter to U.S. EPA, USDA Forest Service, and Lincoln County Health Department. (The same letter was sent to other agencies as well.)

4. Alerted to the problem by telephone, the Lincoln County Health Department initiated its study before the July 17 letter was mailed.

5. Five Rivers residents. August 14, 1979, letter to U.S. EPA Administrator Douglas Costle.

6. Davido, Frank L., U.S. EPA Pesticide Incident Response Officer. August 16, 1979, letter to Mrs. Melyce Connelly, Five Rivers.

7. Wood, Dr. Barbara A., Lincoln County Health Officer. August 29, 1979, letter to Lincoln County Board of Commissioners. Paul Merrell received this document from the Siuslaw National Forest under the Freedom of Information Act.

8. Lincoln County Commissioners. August 30, 1979, letter to U.S. Bureau of Land Management, Siuslaw National Forest, and Oregon State Forestry Department. (Paul Merrell received under FOIA from Siuslaw National Forest.)

9. USDA Siuslaw National Forest. September 4, 1979, press release.

10. Miller, Dr. Charles W., Field Studies Coordinator, Environmental Studies Program, Hazard Evaluation Division (H.E.D.), U.S. EPA October 4, 1979, "Status Report – Five Rivers, Oregon" to Davido, Frank, note 6, op cit. (Paul Merrell received under FOIA from U.S. EPA)

11. Johnson, Edwin L. November 2, 1979, letter to Mrs. Melyce Connelly, Five Rivers.

12. U.S. EPA Howard Johnson Motor Inn meeting, Chapter 12, note 14, op cit., p. 3.

13. Johnson, Edwin L. June 13, 1979, presentation at Forest Industry's Environmental Forum, enclosure to letter to Melyce Connelly, note 11, op cit.

14. U.S. EPA, Data Quality Status Report, Chapter 12, note 27, op cit.

15. U.S. EPA Notice. Federal Register 43(234):57000, December 5, 1978.

16. Health Protection Branch Canada press packet, Chapter 12, note 30, op cit.

17. Reuber, Melvin. "Carcinogenicity of picloram and 2,4-D." Paper presented June 15, 1979, at symposium, "Pesticides, consequences to human health and ecological systems." University of Oregon Health Sciences Center, Portland, Oregon. See also Dr. Reuber's background papers in Appendices volume to U.S. Senate Subcommittee Staff Report, Chapter 2, note 17, op cit. Paul Merrell said that he had received documents under discovery in Merrell v. Block which confirm that neither Dow nor EPA could locate any raw data for the Dow picloram cancer study.

18. USDA Siuslaw National Forest press release, note 9, op cit,

19. Wheeler, Bill, Colorado Pesticide Studies Center. March 10, 1980 memorandum to Mike Delarco, Health Effects Branch, U.S.E.P.A. (Paul Merrell received under FOIA from U.S. EPA)

20. Documents revealing scope of the Five Rivers Health Study included memos, notes 10 and 19, op cit., but especially the October 4, 1979, document: "Contact has been made with all hospitals in the Alsea II Study and their cooperation has been obtained regarding abstracting and submitting information on hospitalized spontaneous abortions for the year 1978. The State Health Department has supplied a computer tape of all live births in the state for 1978. The computer tape contains the same data as the previous computer tape for 1972-—1977. This latter information is presently undergoing programming to obtain the desired data for the areas previously studied. A post-card survey of the physicians serving the study and control areas is being prepared for administration by the Oregon State Health Department to obtain an estimate of the number of spontaneous abortions which the physicians see and treat in their offices as compared with the number of hospitalized spontaneous abortions."

21. Hughey, Ann, and Garino, David P. "Stauffer is granted patent for herbicide that might rival Monsanto Co.'s Roundup." The Wall Street journal, April 8, 1982, p. 8. According to the article, estimated world-wide sales of Roundup are $500 million, which contributed more than a third of Monsanto's per share earnings in 1981.

22. "Succotash [Alliance] labels Ashford tests a sham." Eatonville, Washington Dispatch, September 24, 1980. See also "Spray 2,4-D and see "Mother Jones, February-March, 1981.

23. Five Rivers residents. July 3, 1981 petition to Senator Mark Hatfield.

24. Woolley, Jack, Director, U.S.EPA Office of Congressional Liaison. September 3, 1981 letter to Senator Mark Hatfield.

25. Stringham, Renée, interview with author. See also "Patients lead doctors to herbicide opposition." Salem (Oregon) Statesman-Journal, November 2, 1980. See also CATS Newsletter, Spring, 1978, p. 7.

26. See EPA memoranda, notes 10 and 19, op cit,

27. See, e.g., Field, Barbara, and Kerr, Charles. "Herbicide use and incidence of neural-tube defects." The Lancet, June 23, 1979, pp. 1341-42.

28. See, e.g., Boffey, Philip M. "Herbicides in Vietnam: AAAS study finds widespread devastation." Science 171:43-—47. January 8, 1971. Meselson's Herbicide Assessment Commission called for further study of "disproportionate rise" in incidence of spina bijida and cleft palate. Meselson, Matthew, et al., Chapter 2, note 10, op cit. See also their preliminary report, p. 7; and GI Guinea Pigs, Chapter 2, note 6, op cit., pp. 158-163.

29. GI Guinea Pigs, Chapter 2, note 6, op cit., pp. 158-163.

30. September 13, 1979, petition from fifteen Lincoln County physicians to county commissioners.

31. December 17, 1979, petition from Pacific Communities Hospital Medical Staff to Lincoln County commissioners.

32. Brown, James H., Oregon State Attorney General. July 11, 1980 opinion, citing the Federal Insecticide, Fungicide and Rodenticide Act, and ORS 527-610, 527.730.

33. Measures 9 and 10 on Lincoln County ballot, November 1980. The history of the Lincoln County initiative is more fully developed in: d' Arbois, Findlay. "Industry dollars inundate herbicide debate." NCAP News, Winter, 1981, pp. 30-33.

34. Financial statements of Citizens for Common Sense, filed with the Lincoln County Clerk, listed contributions from such organizations as Georgia-Pacific Corp., Willamette Industries, Weyerhauser, International Paper Corp., Champion International Corp., Davidson Industries, Longview Fibre Co., Crown-Zellerbach, Boise-Cascade, Northwest Christmas Tree Growers Ass'n., Western Helicopters, Publishers Paper (Times-Mirror), Medford Corp., Louisiana Pacific Corp., Pope & Talbot, Inc., and Oregon Agricultural Aviation Ass'n. Salem (Oregon) Statesman-Journal, November 29, and December 6, 1980.

35. For example, Lincoln County Citizens for Common Sense, June 11, 1980, letter to county voters; and October mailer, same organization, received by at least several county residents on October 24, 1980. The Doctors for Facts campaign material was essentially identical to the Citizens for Common Sense material, and bore the CCS return address. On November 3, 1980, voters found a mailer from "Bill Ferguson, M.D." of Lincoln County Citizens for Common Sense in their mailboxes. If Ferguson was a real person, he was neither practicing in, nor was he a resident of, Lincoln County at the time.

36. A January, 1981, Oregonians for Food and Shelter press packet included collected campaign material text from Lincoln County Citizens for Common Sense. Fourteen pages out of the 34-page total were attributed to Dost and Witt.—Both are USDA Cooperative Extension Service specialists.

37. Herman, Steven, December, 1980, interview with author.

38. Monahan, Terry. "Herbicide spending leads Lincoln County final reports.'" Salem (Oregon) Statesman-Journal, December 6, 1980.

39. Merrell, Paul. "The Industrial Bio—Test Caper." Chapter 12, note 9, op cit.

40. USDA Forest Service, Pacific Northwest Region, Siuslaw National Forest, Vegetation Management Environmental Assessment, April 8, 1981, and attached decision notice.

41. Merrell v. J. R. Block, et al. (U.S.D.C. Oregon), Civil No. 81-6138-E, plaintiff's April 15, 1981 complaint.

42. Merrell v. Block, ibid., plaintiff's December 4, 1981, memorandum in support of motion to compel production and to impose sanctions. Through a Freedom of Information Act request, Peter von Stackelberg had learned that the U.S. Army CBW Command had contracted millions of dollars with IBT to conduct studies. For further insight into the chemical military-industrial complex, see Hersh, Seymour M., Chemical and Biological Warfare: America's Hidden Arsenal. New York: Bobbs—Merrill, 1968, especially chapters 8 and 9. See also Retired General J. H. Rothschild's Tomorrow's Weapons (McGraw—Hili, 1964, p. 31) for an interesting view of the "Bugs and Gas Establishment's" relationship with the chemical industry's testing laboratories. Similarly, the government's own National Center for Toxicological Research has been identified as a thinly-disguised cover for the Pine Bluff Arsenal in Arkansas, a major center for production and development of chemical and biological warfare weapons. Seagrave, Sterling. Yellow Rain: A Journey Through the Terror of Chemical Warfare. New York: Evans & Co., 1981, pp. 276-279.

"One Little Helicopter"

1. Lincoln County Road Department, Daily Spray Reports, June 18, 19, 1975. In U.S.EPA FIFRA Consolidated Docket Numbers 415, et al., In re: The Dow Chemical Company, et al. January 22, 1981, exhibits to direct testimony of Mr. Robert G. Heath. U.S.EPA Exhibit 941.

2. Lincoln County Commissioner Jack Postle, who had halted use of herbicides on county roads in 1976, did not run for re-election in 1978. After that year, county roadside spraying resumed; but until 1981, road crews honored residents' NO SPRAY signs. In 1981, the policy was abandoned without public notice.

3. Detzel, Tom. "Breathing Against the Nozzle." Multnomah Monthly, Portland, Oregon, September, 1981, pp. 4-6, 17.

4. "County road crews ignore 'no spray' Signs." Newport (Oregon) Coastal Monitor, May, 1981. "It's unscented." Coastal Monitor, ibid. "Herbicide activist arrested." Newport (Oregon) Lincoln Log, June 2, 1981. "Five Rivers residents block herbicide truck." Salem (Oregon) Statesman-Journal, May 14, 1981.

5. "Oops—wrong truck!" Newport (Oregon) Lincoln Log, May 26, 1981.

6. See, e.g., Seaver, Harmon, "In defense of my organic homestead." Organic Gardening and Farming 24(10):184, October 1977; see also "Spraying foe accused of firing at helicopter." Minneapolis (Minn.) Tribune, August 3, 1976; Coffman, J. "Protester who fired rifle found not guilty." Minneapolis (Minn.) Tribune, November 11, 1976.

7. "Man acquitted of blocking spray." NCAP News, Winter, 1981, p. 13.

8. A report of the Shah's transfer by Evergreen was released on the Pacific Northwest wires of United Press International on March 24, 1980. The story linked Evergreen to the C.I.A. See also Robbins, Christopher. Air America: The Story of the C.I.A. 's Secret Airlines. New York: Putnam's Sons, 1979, pp. 297-298. According to Robbins, Evergreen has absorbed equipment and personnel from a known C.I.A. proprietary airline, Intermountain Aviation. But Evergreen has denied any connection with the C.I.A. See also Rosenberg, Martin. "New Evergreen chief hopes to build stronger financial base for expansion." Salem (Oregon) Statesman-Journal, March 7, 1982.

9. Personal communications to author from Susan Parker and NCAP staff. The flyers were sent to antiherbicide groups and individuals, mailed from the Los Angeles area with no return address.

10. Arnold, Ron. "The politics of environmentalism." Presentation to the Ontario (Canada) Agricultural Conference, January 8, 1980.

11. Worthington, Richard E., USDA Forest Service Pacific Northwest Regional Forester. December 17, 1980, "Integrated Pest Management" memorandum to Forest Supervisors. Worthington required development of law enforcement plans as part of pesticide project work plans "in conjunction with Forest Service special agents." Each plan was to include "an intelligence report on the local situation." See also January 19, 1981, USDA Siuslaw National Forest Timber Staff Officer John O. Hoffman memorandum to district rangers, "Law enforcement planning in pesticide project work plan development." Hoffman told district rangers that they need not develop such plans at the district level because a comprehensive plan was currently being written at the Forest Supervisor's office. See also USDA Idaho Panhandle National Forest Public Information Officer Susan E. Yonts's April 29, 1981, letter to Mr. Bill London, St. Maries, Idaho. London had queried Idaho Forest Service officials on similar plans after reviewing the latter documents from Oregon. Yonts told London the "intelligence report" mentioned in the Worthington memorandum "would include the name of the organization and its leaders, general information about the group (what they support), and any press clippings or copies of statements they've made in public."

12. Quotation transcribed from original videotape interview conducted by Rick and Trish Gibson, Coast News Service (Otter Rock, Oregon) on May 31, 1981. The Gibsons shared their interview with other members of the media, and it quickly spread. See, e.g., KNPT Radio (Newport, Oregon) evening news broadcast, June I, 1981. See also Rosemary, Kristine. "Genetics Brigade claims it destroyed spray 'copter." Salem (Oregon) Statesman-Journal, June 2, 1981.

13. Holzhauer, Charles. "Better PR—An ag chemical need." Oregon Farmer—Stockman, March 5, 1981, p. 3.

14. Hawkes, Carl, Chapter 2, note 4, op cit.

15. Vawter, Dee, interview with author.

16. Untitled photo caption. Newport (Oregon) News-Times, June 17, 1981, p.2.

17. Rosemary, Kristine, note 12, op cit.

18. "Reward possible in helicopter burning." Newport (Oregon) News-TImes, June 10, 1981, p. 1.

Informed Discretion

1. U.S. Senate Subcommittee Staff Report, Chapter 2, note 17, op cit. See also Comptroller General Report, Chapter 12, note 3, op cit.

2. "The act [NEPA] states a whole lot of lofty sentiments .... But there is no law to be found anywhere in the act. Sentiments only—with a bit of staff and eventually a bit of money thrown in. But no law. No criteria identifying precisely what behavior is thought to be harmful and therefore unlawful. There is not even a small step in this direction.":' Lowi, T. The Politics of Disorder, 1971, pp. 179-80. Cited in Orloff, Neil, and Brooks, George. The National Environmental Policy Act: Cases & Materials. Washington, D.C.: Bureau of National Affairs, 1980, p. 15.

3. The National Environmental Policy Act of 1969, as amended. Pub. L. 91-90,42 U.S.C. §§ 4321-4347, January I, 1970, as amended by Pub.L. 94-52, July 5, 1975, and Pub.L. 94-83, August 9, 1975. Section 101(e).

4. See, e.g., Lathan v. Brinegar, 506 F.2d 677,693 (9th Cir., 1974) (en banc).

5. The U. S. Senate Commerce Committee at one time proposed an amendment to FIFRA which would have authorized citizens to bring injunctive suits against certain violators of the Act, and against EPA for failing to perform mandatory duties. S.Rep. 92-70, 92nd Cong., 2d Sess., 3 U.S. Code Cong. & Ad. News, pp. 4093, 4106-4107, 4123-4126 (1972). The proposed amendment was subsequently deleted from the bill by the

House-Senate Conference Committee. Conf. Rep. 92-1540, 92nd Cong., 2d Sess., 3 U.S. Code Cong. & Ad. News, p. 4134 (1972).

6. Union Carbide and Chevron cases, Chapter 12, note 15, op cit.

7. Madison, James. August 4, 1822, letter to WT. Barry. Cited in dissent of Justice William 0. Douglas, EPA v. Mink, 410 U.S. 73, 110-111 (1973).

8. U.S. Senate Subcommittee Staff Report, Chapter 2, note 17, op cit., p.3.

9. Doutt, Richard, Chapter 12, note 10, op cit.

10. Graham, Frank, Jr. Since Silent Spring. Boston: Houghton-Mifflin, 1970, Fawcett edition, p. 135. In the same work, see discussion at pp. 132-35 of F.A.A. investigations into crop dusters' high incidence of accidents, and effects of poisons on pilot performance. The Friends of the Earth (FOE) Washington, D.C., office has developed a file of FAA-sponsored research into pesticide effects on pilots which FOE staff found in FAA files.

11. Dougherty, R.C., and Pietrowska, K. "Screening for negative chemical ionization mass spectrometry for environmental contamination with toxic residues: Application to human urines." Proceedings of the National Academy of Sciences, U.S.A. 73(6): 1777-81. See also Dougherty, R.C, et al. "Chemical environmental agents: Potential human hazards." In McKinney, James D., ed., Environmental Health Chemistry. Ann Arbor: Ann Arbor Scientific, 1981, pp. 263-278. (Sperm density as a function of the presence of specific toxic substances; relates findings to steady decrease in American male fertility since 1950. See also Jansson, Erik. "The impact of hazardous substances upon infertility among men in the United States." Friends of the Earth, Washington, D.C., 1980 white paper.

12. See, e.g., Graham, Frank, note 10, op cit., pp. 119, 143, 239. See also U. S. Senate Subcommittee Staff Report, Chapter 2, note 17, op cit., p. 5; and results from national study, Chapter 11, note 7, op cit.

13. See, e.g., U.S. Air Force documents, Appendix A to this book. See also Graham, Frank, note 10, op cit., p. 133. Graham cites two F.A.A. studies from the early 1960s for a long list of behavioral effects in humans associated with organophosphate pesticides. See also unpublished U. S. Senate Subcommittee staff report, Chapter 2, note 7, op cit., section on toxicity of TCDD and 2,4,5-T, charts which compile lists of human exposure symptoms.

14. Carson, Rachel. Silent Spring. Boston: Houghton-Mifflin, Fawcett Crest edition, 1962, p. 22.

15. Hall, Ross H. Report of the Canadian Environmental Advisory Council. Report No. 10. July, 1981. New Approach to Pest Control in Canada, Canadian Ministry of the Environment, p. iii,

16. Ibid, p. 42.

17. See Carson, Rachel. Testimony before congressional subcommittee. In Graham, Frank, note 10, op cit., p. 91: "I hope this committee will give serious consideration to a much neglected problem—that of the right of the citizen to be secure in his own home against the intrusion of poisons applied by other persons."

18. Jefferson, Thomas. September 28, 1820, letter to Wm. Charles Jarvis.

Carol Van Strum is a writer, bookseller, ruthless editor, Agéd Parent, and seasoned troublemaker. She is the author of *The Oreo File*, *No Margin of Safety*, and *The Politics of Penta*.

She is featured in the award-winning documentary film The People vs. Agent Orange, which is based in large part on the material in *A Bitter Fog*. For her outstanding environmental and social justice work she was awarded the international David Brower Lifetime Achievement Award in 2018.

She lives in rural Oregon on a small farm with a multitude of animals, including Rudy, Gus, and Sadie.

www.ingramcontent.com/pod-product-compliance
Lightning Source LLC
Chambersburg PA
CBHW060307030426
42336CB00011B/967